The Environmental Impact of Sieben Linden Ecovillage

Environmental impact assessment is widely taught and researched, but rarely covers both lifestyle and building construction in a town or neighbourhood. This book provides a broad assessment of the environmental impact of the ecovillage Sieben Linden in Germany.

The ecovillage was founded in 1997 and has a population of over one hundred people. This book shows how raising the awareness of individuals and adopting a consistent way of community living can be environmentally friendly. This applies both to everyday practices and the way the houses in the ecovillage are built. The tools used to measure the impact are Ecological Footprint and Carbon Footprint methodologies, making use of indicators such as Primary Energy Intensity and Global Warming Potential. Despite the difficulties encountered by using standardised methodologies, these research tools provide an overall assessment and have allowed comparisons with selected, similar cases and general values from statistic sources.

This book will be of great use to professionals and scholars in the fields of environmental impact assessment, particularly at the town/district/city level, and of city and ecovillage management. It will particularly appeal to those engaged in a Sustainable Development Goal #11 perspective, as well as environmental policy makers at the local level.

Andrea Bocco is Associate Professor of Architectural Technology at the Politecnico di Torino, Italy. His research interests cover the work and thoughts of unconventional contemporary architects, local development and construction with natural materials.

Martina Gerace graduated in architecture at the Politecnico di Torino (Master's in Architecture Construction City), Italy, in 2017. She is currently specialising in the design of technical installations in buildings.

Susanna Pollini graduated in architecture at the Politecnico di Torino (Master's Degree in Architecture for Sustainable Design), Italy. She is currently engaged in research on two fronts: the assessment of sustainability in architecture and raw earth construction in the developing countries.

The Environmental Impact of Sieben Linden Ecovillage

Andrea Bocco, Martina Gerace
and Susanna Pollini

Routledge
Taylor & Francis Group
LONDON AND NEW YORK

First published 2019
by Routledge
2 Park Square, Milton Park, Abingdon, Oxon OX14 4RN

and by Routledge
52 Vanderbilt Avenue, New York, NY 10017

First issued in paperback 2020

*Routledge is an imprint of the Taylor & Francis Group,
an informa business*

British Library Cataloguing-in-Publication Data
A catalogue record for this book is available from the British Library

Library of Congress Cataloging-in-Publication Data
A catalog record has been requested for this book

ISBN 13; 978-0-367-67032-0 (pbk)
ISBN 13: 978-0-367-14564-4 (hbk)

Typeset in Times New Roman
by Apex CoVantage, LLC

Contents

List of figures   vii
List of tables   ix
Acknowledgements   xi
List of abbreviations   xiii

Introduction   1

1   Sieben Linden ecovillage   5

2   The environmental impact of the Sieben Linden lifestyle   27

3   The environmental impact of the Sieben Linden buildings   55

4   Comparing daily impact and construction impact   75

5   Final remarks, recommendations and perspectives   81

    Index   97

Figures

1.1 Plan of Sieben Linden   13
1.2 Share of time spent in different kinds of spaces   23
2.1 Overall Carbon Footprint   41
2.2 Comparison between Sieben Linden and German average
     carbon footprints   42
2.3 Comparison of consumption by category, 2014–2002   43
2.4 Comparison between Sieben Linden CF in 2002 and 2014   44
2.5 Overall Ecological Footprint   45
2.6 Comparison between Sieben Linden footprint and German
     average footprint   47
2.7 Planet equivalents for Sieben Linden   48
2.8 Comparison between Sieben Linden's EF and other
     communities' EF   50
3.1 Libelle mass, PEI and GWP and Villa Strohbunt mass, PEI
     and GWP   60
3.2 Libelle technical services by mass, PEI and GWP,
     compared to grand totals   64
3.3 Embodied energy and carbon comparisons between the three
     buildings   67
3.4 Comparison between Libelle, Villa Strohbunt and the mean
     values for the most common construction systems   68
3.5 Comparison of Libelle, Villa Strohbunt and
     Wegmann-Gasser EF   72
4.1 Comparison between Sieben Linden footprint and
     German average footprint   76
4.2 Comparison of operational energy and PEI of buildings
     and of operational emissions and GWP of buildings   78

Tables

2.1	Components and variables used in this study	30
2.2	Energy consumption and footprint	32
2.3	Items possession and acquisition	34
2.4	Goods' footprints	34
2.5	Waste footprints	35
2.6	Breakdown of travel by purpose and means of transport	36
2.7	Travel footprints	37
2.8	Food footprints	38
2.9	Built-up land footprint	40
2.10	Overall Carbon Footprint	40
2.11	Comparison between Sieben Linden and German average carbon footprints	42
2.12	Comparison of consumption by category, 2014–2002	43
2.13	Comparison between Sieben Linden CF in 2002 and 2014	44
2.14	Overall Ecological Footprint	45
2.15	Comparison between Sieben Linden footprint and average German footprint	46
2.16	Comparison between Sieben Linden's EF and other communities' EF	49
3.1	Libelle mass, PEI and GWP and Villa Strohbunt mass, PEI and GWP	59
3.2	Mass, PEI and GWP of Libelle technical services	63
3.3	Comparison of Mass, PEI and GWP values of Libelle, Villa Strohbunt and Wegmann-Gasser house	66
3.4	Comparison between Libelle, Villa Strohbunt and the mean values for the most common construction systems	68
3.5	Libelle and Villa Strohbunt EF	71
4.1	Comparison between Sieben Linden footprint and German average footprint	76
4.2	Energy and emissions of buildings	77

Acknowledgements

Simone Contu introduced us to the EF methodology and gave us support in getting acquainted with its mathematics. Mathis Wackernagel of the Global Footprint Network agreed to furnish us with national datasets. Dirk Scharmer kindly provided technical information on the Libelle house. Chris West gave us support in accessing the EUREAPA database. Rajib Sinha of the Royal Technical University of Sweden and Morten Birkved of the Technical University of Denmark kindly reviewed the draft and provided suggestions on how to improve it.

Christoph Strünke was the guardian angel of this whole project and provided every kind of support from data collection to meetings organisation, and acted as the proactive linkage between the research team and Sieben Linden community. The list of people who kindly accepted to record their daily mobility or to have an interview regarding lifestyle patterns would be too long to write down and would virtually cover most of the residents. However, a special mention must be made of Werner Dyck, Iris Kunze, Martin Stengel, Stella Veciana and Michael Würfel for their specialist contributions.

The Open-Access publication of this book was funded by a Politecnico di Torino individual research grant.

Abbreviations

BC	biocapacity
Bed Zed	Beddington zero energy development
BIM	Building information modeling
CF	Carbon Footprint
DBU	Deutsche Bundesstiftung Umwelt
DEFRA	Department for Environment, Food & Rural Affairs (UK)
DESTATIS	Statistisches Bundesamt (Germany)
DHW	domestic hot water
e.G.	eingetragene Genossenschaft
e.V.	eingetragener Verein
EC	Embodied Carbon
EE	Embodied Energy
EF	Ecological Footprint
EIA	Energy Information Administration (USA)
EPDM	ethylene-propylene diene monomer
ES	ecosystem service(s)
EUROSTAT	Statistical Office of the European Union
FASBA	Fachverband Strohballenbau Deutschland e.V.
FÖJ	Freiwillige Ökologische Jahr
GEN	Global Ecovillage Network
GFA	gross floor area
GFN	Global Footprint Network
gha	global hectares
GHG	greenhouse gas
GIA	gross internal area
GWP	Global warming potential
ICE	Inventory of Carbon and Energy
ISO	International Organization for Standardization
LCA	Life cycle assessment
LCAI	Life cycle impact assessment

MHVR	mechanical ventilation heat recovery
PEI	Primary energy intensity
PEX	cross-linked polyethylene
PV	photovoltaic
SDG	Sustainable development goal
SiGe	Siedlungsgenossenschaft Ökodorf e.G.
TAE	Telekommunikations-Anschluss-Einheit
UAE	Universal-Anschluss-Einheit
UBA	Umweltbundesamt (Germany)
WoGe	Wohnungsgenossenschaft Sieben Linden e.G.

Introduction

The present study was born out of an agreement between the Sieben Linden community (Siedlungsgenossenschaft Ökodorf e.G.) and the Politecnico di Torino (DIST), approved on May 9, 2013. The Politecnico team, led by professor Andrea Bocco, intended to perform an overall analysis of the way of living in the ecovillage, including a number of areas among which are agriculture, biodiversity, building, decision-making, diet, economy, energy, land husbandry, etc. The Sieben Linden community, on the other hand, was particularly interested in having a new ecological impact assessment done, a dozen years after that by the University of Kassel (Dangelmeyer et al. 2004).

Therefore, efforts were focussed on the topic, and data collection activities were developed under the coordination and continuous engagement of Sieben Linden's Christoph Strünke (see p. xi), both relying on existing databases and custom-crafting tools such as interviews and questionnaires. Most data were elaborated in 2016–17 by Martina Gerace and Susanna Pollini in the framework of their Master theses in Architecture at the Politecnico which they defended on September 26, 2017. This report is based on the documentation prepared in view of the final presentation of November 23, 2017 to the Sieben Linden community and was further enriched by the results of the discussion with them, the questions raised at the Global Ecovillage Network (GEN) Germany meeting to which it was presented on May 25, 2018, as well as the comments of disclosed and undisclosed reviewers. Finally, constant interaction with editorial staff at Routledge and requirements to comply with legitimate editorial rules helped the text take a (hopefully) decent book shape.

The methodology chosen is twofold:

1   The environmental impact of the ecovillage residents' lifestyle (that is, recurrent activities which are performed on an everyday basis) was

assessed with reference to two methods, the Carbon Footprint (CF) and the Ecological Footprint (EF) (see Chapter 2).

The first is widespread in environmental assessments of products and companies, but it is rarely applied to the analysis of the lifestyle of individuals. However, it seemed challenging to calculate it, as it allows a comparison with a previous study on Sieben Linden's environmental impact (Dangelmeyer et al. 2004). Moreover, CF is actually part of the Ecological Footprint calculation, although the two methodologies use slightly different approaches to accounting.

The latter (EF) appeared an appropriate tool to produce an inclusive picture of the most relevant activities and quantify them in a single unit of measurement that is easy to visualise and communicate. Since the application of this method at a very large scale – that is, to a small entity such as a hundred-something community extending on a few ten hectares – has been tried quite seldom, it seemed to us a stimulating challenge to check its aptitude to describe even minute phenomena like those we were dealing with. The results obtained seem to confirm the appropriateness of the method to the task and their comparability with akin small-size human groups and ecovillages in particular, in spite of obvious simplifications (see Chapter 5.1), inability to describe phenomena not directly affecting ecosystems, and incomplete data libraries. These and other limitations implied by this method (see also Bjørn 2016, Castellani 2012) have been confirmed by this study. A much more complete, and correspondingly much more complex analysis would adopt other methods such as the "absolute environmental sustainability" approach (Bjørn 2015; Nykjær-Brejnrod 2017).

2   The environmental impact of the construction of Sieben Linden buildings (that is, one-off activities aimed at creating items having an indeterminate "service life") was assessed with reference to two basic sustainability indicators ("embodied energy" or PEI and greenhouse gases emission or GWP) and also "translated" into EF terms (see Chapter 3). Since no data could be collected regarding the energy expenditures at the building site, the service life is included in (1) above, and no dismissal can be envisaged (or, at least, when and how it will happen), only the "cradle to gate" phase was accounted for. Also, in this case, we had to rely on not always complete and specific databases, and moreover we excluded recurring to proprietary databases and software; a complete LCA, although within the same boundaries, would have produced richer and more detailed information on the environmental impacts associated with building construction. In spite of such approximations, we believe that also in this field we were able to obtain satisfactory

results which can be compared against similar case studies, and particularly so against other "green" buildings.

Finally, the results obtained though the two methodologies have been merged (Chapter 4) and discussed (Chapter 5). The latter is the final chapter, which includes a few suggestions to contribute to decreasing Sieben Linden's impact on the environment, and mentions questions open to further research. Moreover, in Chapter 5 we also move a few steps back to look at the wider picture and speculate on the societal innovation potential the Sieben Linden model shows, to the benefit of all those local communities that want to reduce their impact, and on the radical changes that are needed to live within a "fair-share" footprint.

In order to make this text accessible to a broad public that might be committed to an eco-sensitive individual or communal lifestyle, we kept technicalities and related jargon to a minimum. However, this book is mainly targeted to professionals and scholars in the fields of environmental impact assessment, particularly at the town/district/city level, and of city and ecovillage management, particularly those engaged in a Sustainable Development Goal #11 ("Sustainable cities and communities") perspective, as well as environmental policy makers at the local level. We hope that in spite of the non-conventional example, useful lessons can be drawn in the field of design, implementation and measurement of sustainability measures.

The case analysed is German, but the theme is relevant worldwide both because of the applicability of this methodology of analysis to virtually any small town, and because of the global extent, and growth, of alternative housing, co-housing and ecovillages. Such communities are usually more concerned than mainstream groups about their environmental impact, and might find here tools to measure their performance.

The present work was directed and edited by Andrea Bocco, who wrote also Chapter 5 and the Introduction. Martina Gerace is the author of Chapter 3, Susanna Pollini of Chapter 2, and they jointly wrote Chapters 1 and 4.

References

Bjørn, Anders; Michael Zwicky, Hauschild, "Introducing carrying capacity-based normalisation in LCA: framework and development of references at midpoint level", *International Journal of Life Cycle Assessment*, 20, 2015, pp. 1005–1018.

Bjørn, Anders et al, "A proposal to measure environmental sustainability in life cycle assessment", *Ecological Indicators*, 63, 2016, p. 1–13. doi:10.1016/j.ecolind.2015.11.046

Castellani, Valentina; Serenella Sala, "Ecological footprint and life cycle assessment in the sustainability assessment of tourism activities", *Ecological Indicators*, 16, 2012, pp. 135–147. doi:10.1016/j.ecolind.2011.08.002

Dangelmeyer, Peter et al. (eds.), *Ergebnisse des Vorhabens Gemeinschaftliche Lebens- und Wirtschaftsweisen und ihre Umweltrelevanz*, Kassel: Wissenschaftliches Zentrum für weltsystemforschung – Universität Kassel, 2004.

Nykjær Brejnrod, Kathrine et al., "The absolute environmental performance of buildings", *Building and environment*, 119, 2017, p. 87-98. doi: 10.1016/j.buildenv. 2017.04.003

1 Sieben Linden ecovillage

Martina Gerace and Susanna Pollini

Sieben Linden Ecovillage is a settlement established in 1997 in the municipality of Beetzendorf in the Altmark, Saxony-Anhalt (Germany). Its aim is to represent a "model of socio-ecological settlement for climate and resource conscious lifestyles and regional development" (Kunze 2016:5). The community's vision and goals are set out in a series of guiding principles, to which new members must adhere and which affect all aspects of life (Kunze 2016:1). Special emphasis is placed on self-sufficiency (especially in food and energy fields), environmental protection and conscious use of natural resources. Sieben Linden is a liberal-minded and hospitable village; it welcomes people from diverse cultural and social backgrounds and age groups, with and without disabilities (Sieben Linden (a)). Sieben Linden is an active member of the GEN[1] and has engaged in an increasing number of cooperative activities and educational programs.

1.1 History

The information used to compile the following section was mainly obtained from Kunze (2016) and Stanellé (2017).

The idea of a self-sufficient ecological village in Germany originated in 1980, during the anti-nuclear resistance in Gorleben. There, an experimental village was built (the "Hüttendorf" of the "Freien Republik Wendland") that lasted for only thirty-three days; however, it was inspirational for many people (Andreas 2012a).

In 1989, Jörg Sommer, a lecturer in psychology at the University of Heidelberg, delineated the essential aspects of a self-sufficient village for 300 people; this was the birth of the concept at the base of Sieben Linden. Sommer spoke of this ideal village as an alternative to the capitalist model:

> For other groups, self-sufficiency is a possibility for withdrawing from society; we, on the other hand, are pursuing the goal of developing an

alternative to the existing industrial and consumer society and therefore to have effects that carry over into society.

(Sommer as quoted in Andreas 2012b:136)

At the time, the main focus was on economic self-sufficiency; however, by 1992 the model idea had expanded to include social and ecological dimensions:

> The model character of the planned village consists of the comprehensive attempt to integrate all spheres of life (home life, work, provision, free time) as part of an ecological circular economy.
>
> (Sommer as quoted in Andreas 2012b:137)

It was an idealistic conception that could not be realised in its purity, but was nevertheless very motivating for many. By 1993, Sommer had left the initiative and the focus of the group shifted to a less radical idea of self-sufficiency (Andreas 2012b).

In 1993 the "Ecovillage housing cooperative" was founded (then renamed "Settlement cooperative" in 1999 and subsequently "Housing cooperative" – see Chapter 1.2); this represented the beginning of the planning phase of the ecovillage. In this phase the guidelines for community living, including spatial planning and development, community organisation as well as ethical aspects, were developed.

In the same year the newly established cooperative bought a "project centre" in Groß Chüden, at that time part of the Chüden municipality and now incorporated in the city of Salzwedel, the capital of the Altmark district. In September a first group of fifteen adult and children pioneers moved there to set up the project and experience communal living at first hand.

In 1996, when the ecovillage had not yet been established, the project team was awarded the "TAT-Orte-Preis" by the German Federal Foundation for the Environment (Deutsche Bundesstiftung Umwelt, DBU for short). The competition rewarded exemplary cultural, ecological, economic and social solutions for underdeveloped regions in East Germany. The jury expressed hope that society at large would benefit from this project: "The exceptional degree of engagement (. . .) which stimulates the region and other environmental education initiatives deserves to be acknowledged. In light of its exemplary nature and its transferability, the proposal is officially awarded and honored" (quoted in Andreas 2012b:137). The ecovillage was awarded again in 2002, for the successful realization of the ecovillage at the Sieben Linden site.

In 1997 the location where to establish the ecovillage was found: the Cooperative bought a farm (consisting of an old building connected to the electrical grid and twenty hectares of forest and agricultural land) near the rural village of Poppau, close to a railway line. The site was selected according to some criteria including, among others, existing infrastructure, access to public transport and affordability of land. In June 1997, even before the approval of the development plan for the area, the pioneers' group moved there, living in trailers.

The land was examined for one year between 1997 and 1998 by an urbanist and a permaculture designer. A global planning process started on the site, based on the methods and the three basic principles of permaculture – care for the earth, care for people, fair share (Holmgren 2002). The initial plan identified residential, public, commercial and natural areas; these definitions have always been intended quite loosely, though.[2]

In 1998 the local municipality approved the development plan for a new rural settlement for 300 people, earmarking a buildable area out of agricultural land; given the extraordinary circumstance, the red-green government of Saxony-Anhalt approved this exception to the law on land use. Moreover, the German law forbids living in trailers; however, the Sieben Linden community obtained the permission to host a maximum of fifty-six trailers, used as temporary housing, until the whole settlement would be completed.

The old farmhouse was then refurbished according to ecological architecture principles, and became the first community building of the Ecovillage: the so-called Regiohaus (see Chapter 1.6.7). This building hosted, and still hosts, the main services such as toilets for people living in the trailers, a community kitchen, a dining room, some offices, a children's room, a guest room and a library.

Since 1999 additional infrastructure was established, such as paths, wells, electricity and telephone lines, reed beds, a pond, open-air amphitheatre and windbreaks. Moreover, a five-hectare piece of overexploited agricultural land was gradually transformed into a valuable vegetable garden.

In 2000 the first residential buildings were built (Nordhaus and Südhaus), which were occupied in November 2000 by twenty people. The growth phase which had just started is still ongoing: many people began to join the community and to move into newly built ecological houses; many subprojects arose, enriching the community's activities. A farm in Poppau was also rented in 1998, which served as a transitional area for those interested in the ecovillage project (which now belongs to an autonomous community closely linked to Sieben Linden).

Since 2010 the educational activities in the ecovillage have been expanding more and more. Educational programs, workshops and seminars attract people from all over Germany; for this reason, in 2013 the Regiohaus was extended with the construction of the Sonneneck, to accommodate guestrooms, toilets, seminar rooms and a reception.

The ecovillage is constantly growing: in 2017 there were eleven multi-family houses, about fifty trailers, the service centre cum seminar space, an organic food shop, an information point, a meditation house, a food storage space, a sauna, several outdoor kitchens and toilets, as well as a carpentry workshop and a horse stable (see Chapter 1.6). A new building is planned opposite to the Regiohaus, with additional bedrooms for guests and spaces for seminars and workshops.

1.2 Organisations in the ecovillage

Sieben Linden is organised in associations and cooperatives. Their aim is to improve the material condition of their members through self-help (Stanellé 2017:29), providing services and also jobs, and managing specific realms of the village's life. This is in accord with the principles of self-sufficiency of the community.

Two cooperatives formally own the ecovillage and manage its finances:

- The Settlement cooperative (Siedlungsgenossenschaft Ökodorf e.G. – in short, SiGe) is the most important organisational unit in the ecovillage. SiGe is the owner of the land and the infrastructure; it decides on all matters that affect the ecovillage and is responsible for managing the economic capital of the community. Joining the cooperative involves obligations towards the community and gives access to all services reserved to residents (under payment of a small fee to cover daily operating costs: see Chapter 1.3). (Kunze 2016:3, Strünke in Stanellé 2017:26)
- The Housing cooperative (Wohnungsgenossenschaft Sieben Linden e.G. – in short, WoGe) is the owner of all buildings. The private property of real estate is not allowed within Sieben Linden; in this way the risk of speculation or acquisition by non-members is avoided and the pursuit of community principles is ensured. WoGe is responsible for the construction and financing of new buildings: when a group of residents decides to build a new house, each individual contributes with a share of money and of working hours; the missing capital is provided by the cooperative. This share will be covered over time through the rent, paid by residents to WoGe. (Kunze 2016:3, Lakas in Stanellé 2017:30)

Another fundamental institution is the Friends of the Ecovillage association (Freundeskreis Ökodorf e.V.). This is a non-profit organisation, open to both residents and non-resident people. It is responsible for public relations, educational programs, cooperation programs, as well as assistance to the youth, the elderly and the disabled. Non-resident members can support eco-village activities with donations and enjoy some advantages (e.g. discounts on seminar and accommodation fees) (Sieben Linden (b)).

Other activities are managed by further associations and groups; among others, the Naturwaren Sieben Linden e.v. supplies its members as well as the guests with natural products such as food, cosmetics and garments (Sieben Linden (c)).

Since its foundation, the ecovillage was organised in sub-communities, called "neighbourhoods" (Nachbarschaften), each with their own conceptual approach and way of life. This structure derived from the idea that groups of residents who share the same values can decide to plan, organise and build a space where to live collectively, pursuing their ideals; in fact, the ecovillage was designed to accommodate the needs of a wide variety of people, assuming a growth of up to 300 residents (Campe in Stanellé 2017:108).

Groups occupy a whole spectrum of approaches to communal living: the radical Club99 group shared the economy and many more intimate things; other groups have more conventional cooperative arrangements; one group came together simply for the practical reason of sharing childcare (World Habitat Awards). People are able to move between the different neighbourhoods if they find themselves more attracted by a different interest group.

The size of a neighbourhood may vary from a minimum of three adults to twenty people. It is not expected that individuals or single families build their own house; in fact, the ecovillage is not just an ecological settlement where everyone could avoid each other if necessary, but a community where they make decisions together and cultivate a sense of community.

New members are integrated into the community through a training path. It consists of a two-week seminar, followed by a one-year trial stay; this allows future residents to directly experience living in the ecovillage and integrate with the community. After the trial period all residents decide about taking the new members into the community (two-thirds of the residents need to approve) (Sieben Linden (d)).

1.3 Work and economy

The information used to compile this section was mainly obtained from Strünke (in Stanellé 2017) and Sieben Linden website (Sieben Linden (e)).

Sieben Linden mainly stands on its residents' extensive labour, both voluntary and paid. Part of the maintenance and expansion activities are performed on a voluntary base: each resident provides regular household services (one to four hours per week, depending on the use of the common areas) and working hours in teams.

Several jobs have been created on-site, many of which are meant to provide services to the community as a whole or the community members as individuals, and therefore substantiate a wholly formalised "circular economy." Currently more than fifty members are employed within the ecovillage; the employers are the Settlement cooperative (forest and garden work, maintenance of the infrastructure), the Housing cooperative (construction of houses, administration), the Friends of the Ecovillage association (educational and seminar business, cooking, cleaning), the Naturwaren e.V. (production and processing of food, organisation of food supply), the Freie Schule Altmark e.V. (nursery), the Rohkostversand "Raw Living" (a company producing vegan foods, most of which are sold to the outside market) and other businesses. Furthermore, there are many self-employed people, who make their work available inside and outside the ecovillage (e.g. seminar leaders, craftsmen, consultants, music and dance teachers, illustrators, etc.).

Some members are employed outside the community, working in fields such as nature conservation, teaching, medicine, music, psychology, social work, seminar facilitation, education and research.

Currently, about 50% of adult residents work inside the ecovillage. Their income is generated either by the internal economic cycle (the cooperatives and individuals) or outside money (especially guests); 30% work outside; 8% receive a pension; 3% are on unemployment benefits; 8% receive other funding (e.g. child allowance, savings, support from the partner). The unemployment rate in Sieben Linden is much lower than in Saxony-Anhalt or in the Altmark, certainly also due to the fact that the eco-village's population is younger than average (thirty-five years, against forty-seven in Saxony-Anhalt).

Living in the ecovillage implies some expenses. Basically, residents are responsible for their own economy, although a neighbourhood may decide to partially or totally share its economy. The entry fee is 1,500 € plus the minimum compulsory shares for membership of the Settlement cooperative (11,275 €) and the Housing cooperative (12,000 €) which are returned if one leaves. In addition, there are annual membership fees for the Naturwaren Sieben Linden e.V. (300 €) and the Freundeskreis eV (80 €). Monthly fixed costs include the use fee for common spaces and infrastructure (about 150 €, 35 € for children), the food fee (190 €), the cooking fee (20 €) and the apartment or trailer rental (variable).

1.4 Self-sufficiency

Self-sufficiency is one of the central pillars of the ecovillage since its conception. From the economic point of view, a step in that direction is done thanks to the work and activities carried out inside the ecovillage. In fact, the money that flows into the ecovillage circulates several times before leaving the village.

In areas such as housing and food, a high level of self-sufficiency has been reached. Our calculations (see Chapter 2) show that in 2014, self-production of electricity (on-site PV panels) was 67% of overall consumption, while 100% of firewood (for heating) was harvested in the community's forest (according to year the self-production actually covers 65–100% of overall consumption), and 61% of water used was extracted from on-site wells (the connection to public aqueduct was recently imposed by bylaw). In the same year, 29% of food was self-produced (that is, 64% of vegetables and 35% of fruits). Moreover, about 74% of the overall consumption was grown in Germany and less than 5% was imported from overseas.

1.5 Rules and guiding principles

The vision and goals of the community are set out in some guiding principles, to which all members must adhere. Moreover, residents are required to participate in community life and activities, including voluntary work and conflict resolution processes.

Guests are expected to respect residents' private spaces (houses and surrounding area) and to conduct themselves in accord with some rules while in the ecovillage (e.g. it is forbidden to use mobile phones and wireless networks inside the ecovillage; it is forbidden to smoke outside the smoking area).

In order to reach a high level of sustainability, the community has drawn up rules in the field of construction, diet and mobility. Some of the most relevant are listed here (Kunze 2016:10–21).

Buildings

- planning of a new house initiated by a group including three adults at the least;
- design by the future inhabitants with the collaboration, if deemed necessary, of professionals;
- private space per capita (excluding common areas of residential buildings) lower than 16 m^2;
- construction process involving future tenants regardless of different ages, education, skills and origin: even children are free to participate;
- lowest possible production of construction waste;
- use of ecological building techniques and materials, which should be as local as possible: timber and earth from the ecovillage's own land, straw from local farmers;
- integration of solar systems for electricity production and water heating;
- heating with wood stoves and use of local firewood;

- low operational energy standards;
- greywater treated in reed bed;
- exclusive use of dry composting toilets.

Diet

- on-site self-production of fruit, vegetables and cereals;
- exclusive use of organic cultivation techniques;
- exclusively vegetarian food served by the community kitchen for guests and residents alike (some dishes in each meal are vegan).

Mobility

- circulation of motor vehicles forbidden within the ecovillage;
- paths covered with gravel, to ensure soil permeability;
- no external night lighting, except for a few LED lamps powered by solar panels near to the Regiohaus and the parking;
- sharing of private cars.

1.6 Spatial description

The information used to compile the following section was mainly obtained from Würfel (2012).

Since its foundation in 1997, the land owned by SiGe has increased from twenty-five to approximately 100 hectares; it includes sixty-four hectares of forest, twenty-five of arable land, six of gardens, and six of buildable areas. In total, 685,000 € were spent to buy the land (Strünke in Stanellé 2017:27).

Currently (2017), about one hundred adults and forty children and teenagers live in thirteen houses and about fifty trailers. The total built surface – including community facilities – is 5,124 m².

Figure 1.1 shows the plan of Sieben Linden with the indication of the buildings and the main areas.

Access to the ecovillage is via a dirt road that crosses it from west to east. Here narrower routes and paths lead to its various areas and neighbourhoods. To the north, there is parking and the joinery workshop, behind which lie the youth neighbourhood, the camping for guests and the forest kindergarten. To the north extends the forest. To the south of the entrance, there is a wide lawn, on which a seminar facility cum guest accommodation building will soon be erected, and the Globolo – the area devoted to spirituality. Proceeding along the main road, one reaches the core of the ecovillage: a former farmhouse, where most communal facilities are found. A square, an amphitheatre and a swimming pond are next to this building,

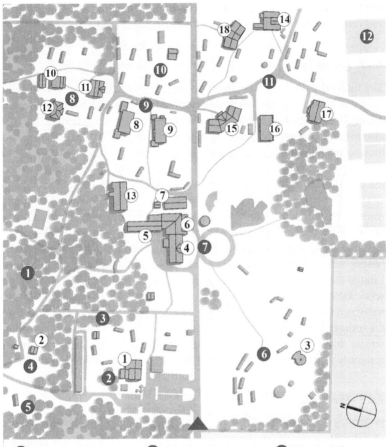

1 Forest   **5** Outdoor kindergarden   **9** District 81⁵

2 Production area   **6** Globolo   **10** South residential area

3 Youth district   **7** Core area   **11** Platz der Unendlichen

4 Camping ground   **8** Nordschonung   **12** Fields

(1) Carperter's workshop   (7) Sauna   (13) Strohopolis

(2) Kubus   (8) Nordhaus   (14) Brunnenwiese

(3) Meditationhaus   (9) Südhaus   (15) Windrose

(4) Regiohaus   (10) Villa Strohbunt   (16) Libelle

(5) Nordriegel   (11) Villa Communia   (17) Nachtigall

(6) Sonneneck   (12) Einhorn   (18) Kranich

Figure 1.1 Plan of Sieben Linden

complementing its functions. To the east are located the residential areas, and further to the southeast extends the agricultural land.

1.6.1 The forest

The forest is a pine and fir plantation created before the community bought the land, which is now being slowly re-naturalised and converted into a mixed forest. It is used to obtain firewood and construction lumber, but it is also meant as an area of peace. A team is in charge of felling the trees and transporting the trunks, with the help of horses, to the production area where they are cut and cleaved; firewood is then conserved under a long shed on the premises or in the many small sheds scattered on the whole ecovillage site. Some Sieben Linden houses were built with timber obtained from this forest. Thus, in some cases, the building materials wood, straw and clay come from a ten-km radius at the most.

In the forest the Kacktempel ("poo temple") is also located – the composting site, which was the first construction in the ecovillage. In Sieben Linden there are only composting toilets which, therefore, do not need a septic tank. Faeces are collected from outdoor as well as indoor toilets and transported with wheelbarrows to this site. This system is more flexible than the ordinary outdoor compost toilet, which constantly requires composting on site, during which time it must be put out of order. The compost produced is mainly used to fertilise hedges, reforested areas and, possibly, orchards.

1.6.2 The production area

In 2003 a carpenter's workshop was built here and equipped with professional machinery, which was expanded in 2010 by adding a new room for further machinery. On the first floor there is a do-it-yourself workshop, financed by donations, which is available to all Sieben Linden inhabitants, and contains tools, work tables, a sink and an industrial sewing machine. Some residents have set up ateliers and a storage, and often the kindergarten performs activities there.

In this zone of the ecovillage there are also storage areas for construction lumber and firewood, and a waste shed. The roof of the firewood shed was the first equipped with photovoltaic panels; the community tractor is also sheltered under this roof. The tractor is used to assist construction activity and to move the caravans around.

The location of these functions in this area of the village was chosen because of the proximity to access road and parking: supplies are facilitated and no means of transport need to progress further. Moreover, noises stay distant from the residential area.

1.6.3 The youth district

This is the area where the young people who participate annually in the ecological volunteer program live (Freiwillige Ökologische Jahr, FÖJ for short). There is a shared straw bale kitchen, and several caravans whose rent is paid by the organisation which employs them. There exists an area specifically allocated to the young people thanks to the suggestion of two former volunteers, now residents, who had found it somewhat difficult to socialise with the rest of the group, as a consequence of the then-random distribution of caravans in different areas of the village.

1.6.4 The camping ground

Until 2008 there were some caravans here but it was decided to move them inside the buildable area and to use this area as a visitors' camp instead. In 2010, a solar shower was added to the existing outdoor toilets.

Besides the camping ground, several accommodation options are available for guests:

• two rooms sleeping four persons each, on the first floor of the Sonneneck;
• a room for three to four persons and two single rooms on the first floor of the Nordriegel;
• a bungalow for five people and a caravan for two, to the east of the production area;
• the Kubus, a small straw bale building located at the edge of the forest in the campsite, which was erected in 2009 during a straw bale workshop. It is mostly passively heated, but a stove is available for additional heating needs;
• the Zelthütte, a small wooden building next to the Globolo. It has a mezzanine, is uninsulated and unheated and is not connected to the electrical grid.

1.6.5 The outdoor kindergarten

West of the campsite, a caravan serves as base for the kindergarten. This was founded in 2001 and is attended by all ecovillage children. At first it was planned a little outside Sieben Linden but it was not possible because hunters were against children roaming in the forest. So it was finally developed on Sieben Linden land. The children spend most of their time outdoors all year round and the entire forest is available for their activities, but they can gather in the caravan to warm up. Close to the caravan hills, caverns, swings and other games have been built. The team consists of three educators plus a guy on his ecological volunteering year (FÖJ). Children attend on weekday mornings.

1.6.6 The Globolo

To the southwest of the village lies this area enclosed by a large circle of robinia wood poles dotted with flowers and herbs. It is a place designed for meditation, music, singing and connection with nature. It sits on a "dragon line," which according to some geomancers passes through Sieben Linden – an energy line that can be compared to a body meridian in traditional Chinese medicine. The area has never really been designed and in fact partially overlaps a site where a second parking area might be built in the future. Within the circle, in summer, yurts are mounted and various activities are carried out inside them. A small building (Meditationshaus) reserved for prayer, meditation and silence has also just been completed here. This structure was built thanks to unpaid work, and funded by donations (Sieben Linden (f)).

1.6.7 The core area

The core area is composed by the L-shaped former farmhouse and two open areas to the north and to the south of it, plus a couple of ancillary buildings to the east.

To the north, an open space that used to be the farmstead's courtyard is to date the most lively spot in the village. At the forest edge, there is a large wooden bench, sheltered by a roof, which is the only place in the ecovillage where smoking is allowed. Adjoining the courtyard are also a small football and volleyball field, a summer outdoor kitchen made of straw bales and a place where residents drop what they do not use anymore, to be freely picked up by others.

The L-shaped building goes under three different names: Regiohaus, Nordriegel and Sonneneck. All together, they constitute the core of the ecovillage and include most of the communal spaces. In 1998–99, with the help of travelling artisans and volunteers, the Regiohaus was refurbished and extended, the timber frame was strengthened, and it was equipped with a new roof. The building was insulated with cellulose flakes and clad in larch boards. It was meant as the community building and has become the focal point of ecovillage life. The name was chosen to imply that this is also a meeting place for the people of the region. In fact, it is the main meeting point for visitors, and contains a small exhibition on the ecovillage and information on scheduled events. On the ground floor there are toilets, a library, a playroom for children, two kitchens and a dining room. The first floor contains a few guestrooms and a large room used for seminars and internal meetings.

The other wing of the old farmhouse is called Nordriegel, as it lies to the north-east of the Regiohaus. The upper story contains guestrooms, community rooms and the offices of most of the community organisations, while at the ground floor the workshops, information office, and bar face the courtyard. A ballroom was built in 2011 as an extension of the bar and is used for dancing on Saturday evenings and for courses during the rest of the week. To the east of Nordriegel there is also a shed for food storage and a sauna. The latter was built by the Club99 with the collaboration of young volunteers as a gift to the whole ecovillage (Kommerell in Stenellé 2017:84).

The Sonneneck is sandwiched between the Regiohaus and the Nordriegel, at the intersection of the two wings. It was built in 2013, *en lieu* of an old workshop. It was designed to provide further spaces for seminars and a dining room for their participants. On the ground floor there is a natural products shop run by the Siedlungsgenossenschaft, open to both Sieben Linden residents and visitors. The name Sonneneck is due to the generous terrace upstairs, a cherished resting place whenever the sun shines.

To the south of the Regiohaus there is a circular space that is meant to become the village square. However, except for a small info pavilion, the square is presently undeveloped and therefore opens to the south towards a large lawn and allows a view of Poppau in the distance. Next to the square lie the amphitheatre, where various activities take place in the summer, and the pond that, in addition to providing a water reserve in case of fire, is a place for fun both in summer, for swimming, and in winter, to skate on ice.

1.6.8 District 81[5]

This is one of the first neighbourhoods to have been built in Sieben Linden (2000–01). It consists of two houses: the Südhaus, which actually consists of three terraced flats, and the Nordhaus which is a large shared house with three kitchens. The name is due to the base dimension of the timber frame (81.5 cm). The initiators of these buildings were united by the desire to live in a real house (as opposed to a trailer), in many cases because of having young children. So, this neighbourhood is focussed on living spaces for families, without ideological implications.

1.6.9 The Nordschonung

The idea of the Club99 had its origins before the Sieben Linden land was bought, even though this neighbourhood was officially founded on

September 9, 1999. Members wanted to experiment with an alternative life-style, based on the following principles (Wiegand et al. 2006:11):

- deep connection with nature, considering all living beings and the quality of soil, water and air;
- engagement to understand the meaning and purpose of human life on this planet;
- collective and individual growth based on social relations, non-violent communication and happiness;
- a wholly vegan lifestyle, which implies not exploiting the animals besides not taking their lives;
- avoidance of products obtained through the exploitation of human beings;
- meeting all needs using renewable and regional resources only;
- income sharing.

The Club99 settled at the north-east corner of the buildable area, in a place called Nordschonung. The buildings in this neighbourhood are Villa Strohbunt, two domes attached to each other (Strohballenkuppeln), and Villa Communia; in addition, there are several trailers, and a house called Einhorn which has just been completed.

The construction of Villa Strohbunt (see Chapter 3.1) began in 2001 and ended in 2004; it was completely built by hand and with local materials and was one of the first straw bale buildings in Germany. It was designed as a common space, but after the closure of the Club99 it was converted into a dwelling.

The two domes north of Villa Strohbunt were built in 2003; one is a guest room, the other a bathroom with two bathtubs and a stove. The domes are made of load-bearing straw bales, like igloos, and plastered with clay. A freestanding roof protects them from the weather.

In 2007 the Villa Communia was built with some compromises with respect to the principles of Villa Strohbunt; some machinery was used, and a telephone line and power sockets were installed. At the time Villa Communia was erected, the building technique had evolved thanks to the fact that ecovillage had become a centre of experimentation for straw bale building. So it has a wooden frame infilled with straw bales, as opposed to Villa Strohbunt where the loadbearing structure is detached from the insulating envelope. At the beginning, Villa Communia did not include a bathroom, but finally one was created at the request of the inhabitants.

Over the years many people have come to Sieben Linden attracted by the Club99, but having difficulty in adapting to its rigid principles have finally settled in other neighbourhoods. For this and other reasons, Club99 failed to grow and, in 2010, it was decided to end the experiment. However, the closure of pioneering projects, namely the Club99, represented a shift from the founding principles of the ecovillage to a less radical lifestyle.

In 2017 another building was completed in this area: Einhorn, a three-storey house that accommodates nine people (Deltagrün Architektur).

1.6.10 Strohpolis

Strohpolis is a large, three-storeyed residential building, with four apartments and three single rooms. The two upper storeys consist of two large apartments with seven bedrooms, a large living room–kitchen and two bathrooms. The large apartments can easily be divided into two apartments with three bedrooms and a kitchen/living room each, or into a five-bedroom apartment and another with a bedroom, depending on the inhabitants' needs. Presently, it hosts nine children and twelve adults.

The WoGe decided to build it to facilitate the relocation of families with young children; it was designed having new ecovillage members in mind, to help them get to know each other, form a neighbourhood group, and possibly leave after a while to start new neighbourhoods. (Actually, only one woman left Strohpolis to settle in a new house.)

Built between 2004 and 2005 with the help of more than 240 volunteers, it was in its time the largest straw-bale building in Europe. It has a timber frame structure insulated with straw bales, plastered with clay (Wiegand et al. 2006:14–18).

1.6.11 Caravans

Living in caravans is meant as transitory, both to save energy and because the ecovillage is based on shared living. Currently around 40% of the residents live in trailers, in the wake of the construction of new houses; it is planned that in the future they will reduce to 10%. The three main areas for such temporary housing are north of the Globolo, the southeast field and the Platz der Unendlichen ("place of the boundless people"). The trailers are all different from one another and have been constantly modified in time: some have added a terrace, others have a closed porch, others have raised their trailer.

1.6.12 South residential area

In this area are all the remainder houses: in chronological order, the Brunnenwiese, the Windrose, the Libelle, the Nachtigall and the Kranich.

Brunnenwiese

This is the first of three houses designed for the neighbourhood to which the caravans on the adjacent lawn also belong. Brunnenwiese is therefore not only a house but also a neighbourhood, whose shared key points are: awareness, spirituality and attention, ecology, homeopathy, meditation, yoga and the creation of a family.

The house was partly self-built in 2004 by prospective tenants and about 150 volunteers. Moreover, the building costs were only partially covered by the WoGe, the remaining share being provided by future inhabitants. The shape is reminiscent of a spiral, at whose centre is a large tree trunk which goes through the "hot room" on the ground floor and the meditation room on the first floor. The kidney-shaped "hot room" is about 10 m² large and is heated by a masonry stove which retains heat in the mass of its raw earth walls. From here, hot air is distributed throughout the house; the temperature in the rooms depends on whether the doors are open and their distance from this warm core. In winter, the "hot room" is also used as a sauna, and also to dry clothes, herbs and fruit. Residents also regard it as a place to share emotions. As opposed to the "hot room," the meditation room upstairs is a very bright space from which one enjoys a broad view of the fields. Aside from these special rooms, the house has seven bedrooms, a fully glazed kitchen-living room, a bathroom and a toilet. The design concept was in fact to create spaces for sharing and communication, and private rooms where one could withdraw.

Brunnenwiese has a timber structure, which on the ground floor is insulated with hemp and clad in timber boards both inside and outside, while the upper-floor walls are infilled with clay-rendered straw bales (the parts most exposed to the weather are also timber-clad). Timber was obtained from the village's forest and the clay from next to the house itself, while the straw bales came from an organic farm close by and hemp from regional crops (Wiegand et al. 2006:18–21).

Windrose

This building was built between 2008 and 2009 as a co-housing for fifteen people (nine adults and six children). However, it emerged that not all inhabitants had the same expectations about shared life: some wanted a more private space while others a more shared one. For this reason, after three years a new private kitchen was created, at the expense of a bedroom.

The house consists of prefabricated elements, a further step forward in the construction system in straw bales in the ecovillage. A warehouse was rented to assemble the wooden frames, compress the straw bales inside them, and finally plaster them. These were then assembled in two weeks at the building site to make the walls, which were then covered with a roof. Due to noise problems in other houses, great attention has been paid to sound insulation, introducing doors between the rooms and oversizing the corridors.

Libelle

This house was completed in 2012. It is located south of the Windrose and can accommodate ten people. The south façade is entirely glazed and is topped by a large solar panel array (see Chapter 3.1).

Nachtigall

This building is located south of the Libelle with a view of the fields. It was built in 2014 and contains two apartments: one inhabited by a large family and the other by a small residential community (Sieben Linden (g)).

Kranich

This building is also made of timber and straw; the north and west elevations are clad with larch boards, while south and east façades are rendered (Stroh Unlimited). It was built in 2015 and can accommodate ten people. The western half was designed for a family while the ground floor of the eastern half is accessible to the disabled (Sieben Linden (h)).

1.6.13 The fields

To the south of the residential area lies some of the ecovillage's agricultural land, where in addition to the vegetable gardens are an orchard, several greenhouses and a reed bed. Most of the land is used for the sustenance of the ecovillage, while to the west small gardens are individually tended by some residents.

Not all of the ecovillage farmland is yet used by the community – some fields are rented, others are left fallow to ensure their transition from conventional to organic farming. The agricultural land is managed by the SiGe, who is also in charge of planning the farming, storing the produce and calculating the daily fee residents should pay for eating the food.

The gardens are irrigated making use of greywater purified in the reed bed.

To the south of the fields lies the so-called "new forest" which was created as a compensation for the area built in the ecovillage. Here is also the "bees' house" – a small caravan used as an apiary – and, next to it, the horse stable.

Finally, further south there is a meadow, which was initially used as a pasture for horses but actually not suitable for this function because of the presence of *Jacobaea vulgaris*, a plant harmful to these animals.

1.7 Use of space

The first settlers lived in trailers and shared outdoor facilities, as only a run-down farmhouse existed. The first interventions were focused on refurbishing it and transforming it into a community building: first the wing called Regiohaus (1998–99) then the Nordriegel (2000–02). During those early years there was a growth in the per capita availability of community space compared to residential space: this was due to the low number of residents and the absence of residential buildings. The values tended to coincide from 2001 to 2005, with the completion of the first residential buildings (Nordhaus and Südhaus) and the supply of new community spaces, such as Villa Strohbunt, the Strohballenkuppeln and the joiner's workshop. Later the trend was reversed, as residential spaces began to grow more than facilities. The total floor area per inhabitant grew from 34 m^2/person (1999), of which 21 m^2 was community and 13 m^2 residential space, to 38 m^2/person (2016), including 11 m^2 community and 27 m^2 residential space.

Actually, the ecovillage's buildings are almost never made up of private apartments, but individual rooms and spaces shared between all tenants or a subgroup of these. The spaces of residential buildings were classified as: shared (accessible by all the inhabitants of the building and open to group use), individual (space for exclusive use of one inhabitant), and service spaces (circulation, storage, bathrooms, etc.). According to this classification the total residential area (3,754 m^2) can be broken down in 2,091 m^2 for individual use (including trailers), 812 m^2 for shared use and 851 m^2 for services. This translates into 15 m^2/person for individual use, 6 m^2/person shared and 6 m^2/person for services.

The residents' routine movements within Sieben Linden were investigated through interviews which produced information regarding eleven persons (eight adults and three minors). Although not statistically representative, these might help understand where inhabitants spend their time. Spaces were divided into four categories: own house, somebody else's house (within Sieben Linden), community spaces (including facilities and outdoor spaces), and outside Sieben Linden.

Figure 1.2 shows that interviewees spend most of their time at home (60%), while for the rest of the day they use the community spaces (20%) more than other private homes (7%). Time spent outside of the village is just 13%.

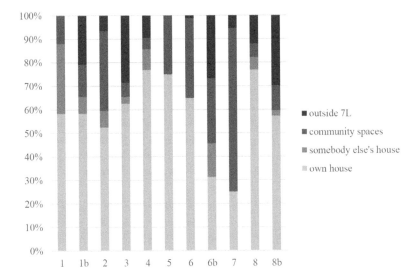

Figure 1.2 Share of time spent in different kinds of spaces

1.8 Relations with the outside world

Sieben Linden aims to represent a practical model of a sustainable community. Communication and exchange of experience are therefore central. Every year the community welcomes many visitors, sometimes for several consecutive weeks (more than 5,000 seminar guest nights per year, plus about 2,000 day visitors (Stützel in Stanellé 2017:118)). The Friends of the Ecovillage association annually organises dozens of events, such as educational seminars on alternative agricultural techniques, theoretical and practical workshops on timber and straw construction, etc.; moreover, once a month the Sunday café offers an opportunity to visitors to get acquainted with the ecovillage through a guided tour. A summer camp takes place yearly (Sieben Linden (d)).

In addition, volunteers from the voluntary ecological year (FÖJ), are annually hosted to help in the everyday routine activities: from 1993 to 2017, about 100 young volunteers have spent a year in Sieben Linden (Strünke in Stanellé 2017:27).

In 2017 a research institute was founded in Sieben Linden. Its aim is to promote particularly relevant research topics and projects for intentional communities in German-speaking countries (Veciana in Stanellé 2017:131).

Contact with the region has also intensified over the years. Sieben Linden is a well-integrated part of the municipality of Beetzendorf and one of its members is a town councillor sitting for the "Energiewendeliste" (energy transition list), which the ecovillage co-founded.

Thanks to the work of the community and the Fachverband Strohballenbau Deutschland e.V. (the German professional association of straw bale construction, FASBA for short), founded in 2002 in Sieben Linden, in 2006 the German government approved the use of (non-load-bearing) straw in buildings. FASBA is a non-profit organisation that performs research, collects and disseminates knowledge, and networks on straw building. The association has now grown to 150 members – artisans, designers and enthusiasts – and has two branches, one in Lüneburg and one in Sieben Linden. Since 2003, FASBA has been involved in research and development projects supported by the Federal Ministry of Food, Agriculture and Consumer Protection and the DBU (Wiegand et al. 2006:4).

In addition, the ecovillage maintains relations and knowledge exchanges at the regional, national and international levels, and has published five editions (the last in 2014) of *Eurotopia*, a directory that collects information on over 400 European intentional communities. Last but not least, from 2004 to 2015 Sieben Linden hosted the European Coordination Office of the GEN.

Notes

1 The Global Ecovillage Network (GEN) is a growing network of regenerative communities and initiatives that bridges cultures, countries and continents. GEN builds bridges between policy makers, governments, NGOs, academics, entrepreneurs, activists, community networks and ecologically minded individuals across the globe in order to develop strategies for a global transition to resilient communities and cultures (GEN).
2 The planning group, in fact, continually revises the original plans to integrate new projects. So there is no master plan to implement; rather the settlement grows and changes over time, as some people leave and others join.

References

Andreas, Marcus, "The ecovillage of Sieben Linden", *Arcadia*, 15, Rachel Carson Center for Environment and Society, 2012a. doi:10.5282/rcc/3917
Andreas, Marcus; Felix Wagner (eds.), *Realizing Utopia: Ecovillages Endeavors and Academic Approaches*, München: Rachel Carson Center for Environment and Society, 2012b.
Deltagrün Architektur, *Familienhaus Einhorn* [online]. Available from: www.deltagruen.de/2017-haus-einhorn/ [last viewed Sept. 2018].
GEN, *About GEN* [online]. Available from: https://ecovillage.org/global-ecovillage-network/about-gen/ [last viewed Sept. 2018].
Holmgren, David, *Permaculture: Principles and Pathways Beyond Sustainability*, Hepburn: Holmgren Design Services, 2002.

Kunze, Iris; Sabine Hielscher, *Fallstudienbericht COSIMA: Entwicklung der Klimaschutzinitiativen*, Poppau: Ökodorf Sieben Linden, 2016. Available from: www.lehmhausen.de/wp-content/uploads/2017/02/Fallstudienbericht_7Linden-TU-Wien.pdf [last viewed Sept. 2018].

Sieben Linden (a), *Sieben Linden Ecovillage* [online]. Available from: https://siebenlinden.org/en/ecovillage-2/sieben-linden/ [last viewed Sept. 2018].

Sieben Linden (b), *Ökonomie: Vereine* [online]. Available from: https://siebenlinden.org/de/oekodorf/oekonomie/vereine/ [last viewed Sept. 2018].

Sieben Linden (c), *Naturwaren Sieben Linden e.V. – Die Foodcoop und der Bioladen* [online]. Available from: https://siebenlinden.org/de/naturwaren-sieben-linden-e-v-die-foodcoop-und-der-bioladen/ [last viewed Sept. 2018].

Sieben Linden (d), *Visit us* [online]. Available from: https://siebenlinden.org/en/visit-us/ [last viewed Sept. 2018].

Sieben Linden (e), *Economy* [online]. Available from: https://siebenlinden.org/en/economy/ [last viewed Sept. 2018].

Sieben Linden (f), *Aktuelle Projekte, Meditationshaus* [online]. Available from: https://siebenlinden.org/de/aktuelle-projekte/meditationshaus/ [last viewed Sept. 2018].

Sieben Linden (g), *Newsletter* [online]. Available from: www.siebenlinden.de/newsletter/DE/no_01_14/ [last viewed Sept. 2018].

Sieben Linden (h), *Freundeskreis Ökodorf – Newsletter* [online]. Available from: http://siebenlinden.de/newsletter/DE/no_04_14/2014-4_newsletter_FK_Oekodorf.html [last viewed Sept. 2018].

Stanellé, Chironya; Iris Kunze (eds.), *20 Jahre Ökodorf Sieben Linden*, Poppau: Freundeskreis Ökodorf, 2017.

Stroh Unlimited, *Wohnhaus Ökodorf Siebenlinden (2015/16)* [online]. Available from: www.stroh-unlimited.de/details/7linden.htm [last viewed Sept. 2018].

Wiegand, Elke; Martin Stengel; Dirk Scharmer, *Ecovillage Sieben Linden with Straw bale construction*, Paper compiled as an entry to the World Habitat Awards 2006 Building and Social Housing Foundation competition, Sieben Linden, 2006.

World Habitat Awards, *Straw-bale Housing in the Sieben Linden Ecovillage* [online]. Available from: www.world-habitat.org/world-habitat-awards/winners-and-finalists/straw-bale-housing-in-the-sieben-linden-ecovillage/ [last viewed Sept. 2018].

Würfel, Michael, *Dorf ohne Kirche. Die ganz grosse Führung durch das Ökodorf Sieben Linden*, Poppau: Eurotopia-Buchversand, 2012.

2 The environmental impact of the Sieben Linden lifestyle

Susanna Pollini

2.1 Methodologies

Different environmental impact accounting methodologies exist; however, just a few are able to account for the whole impact of a specific lifestyle. The methodology we consider as most appropriate is the Ecological Footprint, as it allows to consider a great variety of actions and quantify them in a single unit of measurement that is easy to visualise and communicate.

Another relevant methodology is the Carbon Footprint, which is widespread in environmental analyses. Although it covers just one of the dimensions of environmental impact, we decide to use it as well not least because it allows comparisons with a previous study on the same ecovillage (Dangelmeyer et al. 2004).

The Carbon Footprint is actually included in the Ecological Footprint calculation, however both methodologies use a slightly different approach to accounting; of course, final results are expressed in different units of measurement.

In the following sections a brief description of each methodology will be provided. Differences and similarities will be highlighted, as well as the arrangements made to align the two methodologies.

2.1.1 Carbon Footprint

The Carbon Footprint (CF) is an estimate of the climate change impact of an activity or product. The phrase "carbon footprint" has gained increasing popularity in recent years and is widely used in scientific literature. However, a large range of definitions exist. In general, differences are primarily focused on two key issues: the units of measurement (kilograms of carbon dioxide ($kgCO_2$) or a range of greenhouse gases (GHGs), expressed in kilograms of CO_2-equivalent ($kgCO_{2eq}$)); and the boundaries of the study (only direct, or both direct and indirect emissions) (Wiedmann 2008).

The definition that we considered as the most consistent with the approach of this study is that provided by the Carbon Trust:[1]

A Carbon Footprint measures the total greenhouse gas emissions caused directly and indirectly by a person, organisation, event or product. A carbon footprint is measured in tonnes of carbon dioxide equivalent (tCO_{2eq}).

(Carbon Trust (a))

On the other hand, the CF definition applied in the Ecological Footprint (EF) calculation (see Chapter 2.1.2) takes a slightly different approach; in fact, it only includes CO_2 emissions (instead of a range of GHGs) derived from the combustion of fossil fuels (while emissions from the combustion of biotic energy sources are not accounted for) (GFN (a)).

As the CF is preparatory to the calculation of the EF, some considerations are needed:

- we used both data provided as CO_2 or CO_{2eq}, depending on the available data – in most cases, CO_{2eq}. We assume this choice is acceptable since CO_2 is by far the largest component of emissions produced by industrial processes;
- we revised the CF results before using them in the EF calculation, in order to exclude the emissions derived from biotic sources.

To facilitate data merging, in the calculation of CF we used the same consumption categories ("components") as in EF (see Table 2.1).

As the calculation of emissions associated to a product or activity has to account for a wide variety of data, in some cases we referred to existing studies, e.g. LCAs of products (for further details see Chapter 2.4).

2.1.2 *Ecological Footprint*

Conceived in 1990 by Mathis Wackernagel and William Rees at the University of British Columbia, the EF is now widely used by scientists, businesses, governments, individuals and institutions working to monitor ecological resource use.

The Ecological Footprint (EF) is a means of measuring the environmental impact of everyday activities. It expresses how large an area of biologically productive land and water an individual, population or activity requires to produce all the resources they consume, and to absorb the waste they generate over a one-year time span, using ordinary technology and resource management practices (GFN (a)). By definition, the method accounts for energy and material flows on a yearly basis, and does not include amortisation accountancy of previously generated stocks: activities and material goods are only and wholly accounted in the year when they occur or are produced. The EF tracks the use of six categories of productive areas: cropland, grazing land, fishing grounds (sea), built-up land, forest area and carbon demand on land (energy land or Carbon Footprint). For further information see Wackernagel et al. (2000).

The productive area needed to grow raw materials is calculated in accordance with yield factors; these indicate the amount of regenerated primary product that humans are able to extract per area unit of biologically productive land or water. Only the use of renewable resources for which the planet has bioproductive capacity is accounted for. World-average yield factors were obtained from the Global Footprint Network (GFN).[2] The EF, as it is measured using global average yields, is then normalised by applying equivalence factors. These are multipliers which adjust different land and sea types according to their relative bioproductivity. The equivalence factors are annually updated by the GFN; we used those published in the Global Footprint Network National Footprint Accounts, 2016 Edition (Lin et al. 2016). The final result is expressed in conventional units called global hectares. EF can be compared against biocapacity (BC), which measures the bioproductive supply. The mathematical difference between BC and EF is called either reserve or deficit. (Galli et al. 2007).

Each time methodological improvements are implemented a new edition of the National Footprint Accounts is released and the GFN calculates new national and world average Ecological Footprint and biocapacity values. In this study we refer to the Ecological Footprint accounting guidelines as of 2016 (Lin et al. 2016) and to the current operational standards (Kitzes 2009), which are the Ecological Footprint Standards 2009.

The EF method is applied to study resource demand at a range of scales from the global and national scales down to regions, cities, households or products. The EF of a city or country is simply the sum total of the EF of all the residents of that city or country.

Two approaches to EF accounting exist: the compound and the component method. The main difference is that they draw upon different data sources to estimate the EF. The compound method estimates consumption based on national trade statistics and energy budgets (a "top-down" approach). This methodology is used in the study of the EF of a country. The component method estimates consumption through analyses of material flows and activity components (a "bottom-up" approach). The main sources of data for the component method are local investigations and life cycle studies; the quality of the analysis relies on access to significant databases of environmental information (Lewan 2001:12). The component method is frequently used in studies of sub-national areas. This method is not yet standardised and the

Table 2.1 Components and variables used in this study

Component	Variable	Unit of measurement
energy	electricity	kWh
	PV panels	m^2
	firewood	stacked m^3
	solar panels	m^2
	propane gas	kg
items	non-food products	pcs
	wood products	pcs
waste	recycled paper	m^3
	recycled metal	m^3
	recycled plastic	m^3
	recycled electronics	m^3
	mixed waste	m^3
travel	travel by car	km/person
	travel by bus	km/person
	travel by urban public transportation	km/person
	travel by airplane	km/person
	travel by ship	km/person
food	food	kg
other	built-up land	m^2
	services	–

results of different studies vary so much that they cannot easily compare (see Chapter 2.5.4). A template method should be developed, as it is suggested in the report presented at the workshop on Ecological Footprints of sub-national geographical areas held in Oslo in August 2012 (Lewan 2001).

The present study follows the component method; Table 2.1 shows the variables used.

2.2 Boundaries of study and functional units

A fundamental question is whether the aim of the study is to assess the footprint of the ecovillage or of its community. The first approach (geographical approach) just considers the activities carried out inside the physical boundary of the ecovillage; the second (responsibility principle) accounts for the consumption of the ecovillage's residents, independently of its physical boundaries (Lewan 2001:4). In the present study, the latter approach has been adopted.

The functional unit is the consumption of resources (energy and materials) and the production of waste of Sieben Linden community in 2014 (observation year), considered as a whole and without distinctions according to age, status, gender or other categories. At that time, 130 people lived in the ecovillage (98 adults, 32 minors); this figure includes full members, people in a trial year, long-term ecological volunteers from FÖJ and private guests. Three residents (adults) left the ecovillage during that year; they are not considered in the study.

In both CF and EF calculations we only accounted for activities carried out and material goods bought in the observation year, 2014 (as required by EF methodology); no amortisation to spread the impacts over a hypothesised life cycle of products has been performed.

2.3 Data collection and revision

Data were derived from a variety of sources. The ecovillage provided detailed information about the community's activities, and specifically about resource consumption and waste production, over the observation year, or exceptionally over a one-year period between 2014 and 2015. Data not covered otherwise were collected through interviews.

In order to avoid accounting errors, all the data have been reviewed before use in consideration of what follows:

• Sieben Linden runs several seminars every year. These attract people who stay for a few days or even weeks. Data for energy, waste and food components cannot but include both residents and guests. In order to account for the consumption by the residents only, the guests' share has been subtracted. The latter has been calculated on the basis of the number of nights spent by guests in the ecovillage, and other specific information provided by the community.
• On the other hand, data for energy, waste and food only refer to the consumption of residents within the ecovillage. On average, residents spent 273 days in Sieben Linden. Annual consumption has therefore been calculated by multiplying the daily consumption by 365. This might imply some underestimation as 1) it is possible that lifestyle patterns are not as virtuous outside the village as they can be in the village; 2) minors, who must have a slightly lower EF, show a tendency to spend more days (300) in the village than adults (264).

2.4 Footprint calculations

2.4.1 Energy

This section includes household energy use from electricity, heating and cooking.

Table 2.2 Energy consumption and footprint

Energy type and source	Use [kWh/person]	CF [kgCO$_{2eq}$/person]	EF [gha/person]		
			energy land	forest land	total
electricity					
grid (EO.N)	356.03	0.00	0.000	0.000	*0.000*
PV panels	294.10	0.00	0.000	0.000	*0.000*
heating					
firewood	3,209.51	192.57	0.000	0.557	*0.557*
solar panels	600.00	24.13	0.006	0.000	*0.006*
propane gas	311.57	67.05	0.017	0.000	*0.017*
total	*4,771.20*	*283.75*	*0.023*	*0.557*	*0.581*

Data on electricity use were derived from meter readings; the ecovillage only uses "green electricity" 100% derived from renewable sources (EO.N-Ökostromprodukte and on site photovoltaic panels). Firewood is used for space and water heating; data on firewood consumption were provided by the ecovillage. In 2014 the ecovillage only burnt firewood obtained from its own forest; the forest is managed sustainably and no more firewood is extracted than the forest's annual growth. Sanitary water is also heated by solar panels, which produce about 600 kWh/person; in 2014, 16 m^2 of new solar panels have been installed. Data on propane gas consumption for cooking are derived from bills. These cover both private kitchens and the community kitchen. Information on firewood and propane consumption were provided in different units and converted in kWh using the calorific value of each fuel.[3]

Table 2.2 shows a breakdown of the energy used and the corresponding footprints.

Carbon Footprint

The total energy CF is 283.75 kgCO$_{2eq}$/person. Electricity has no impact: the impact of on-site PV panels has not been accounted as no new one had been installed in 2014, the observation year for this study; electricity purchased from the power grid has been assessed as zero impact as it is 100% generated from renewable sources, while power stations and transmission infrastructure are accounted in the services (see Chapter 2.4.7). Heating is the greatest contributor to the footprint, with 76% of total CF; emissions from firewood combustion represent almost 90% of this share (data on firewood emissions were extracted from Francescato et al. (2004)), while the remaining part is due to the production of solar panels installed in 2014 (data from Menzies and Roderick (2010)). Propane gas used for cooking only represents 24% of the CF (data on propane emissions from EIA (2010)).

Ecological footprint

Total energy EF is 0.581 gha/person. More than 96% of the footprint is represented by the forest area needed to produce firewood; as is typical of this method, EF tends to emphasise the impacts connected to biotic productions. The remaining part of the EF is represented by the forest area needed to absorb the CO_2 emission derived from the combustion of propane gas and the production of the solar panels installed in 2014; no land is accounted to absorb the emissions derived from firewood combustion (see Chapter 2.1.1).

2.4.2 Goods

Data for the goods category were available for a selection of durable products (vehicles, office equipment, home appliances) and non-durable products (clothes, books, magazines, newspapers). The first were derived from an inventory of all such items existing in Sieben Linden; in accordance to EF methodology, only new items bought in 2014 were accounted in the study. Data on non-durable products were based on interviews, which for instance revealed an average acquisition of eight new garments, seven new books, and fifteen magazines per year and per capita, and 0.2 newspapers per day and per capita. Per capita 0.8 mobile telephones are possessed (50% are smartphones), in spite of the prohibition of making use of them within the ecovillage. On average, Sieben Linden residents spend five hours on the Internet per capita and per day.

Table 2.3 shows items possession and acquisition in Sieben Linden from 1996 to 2016, while Table 2.4 shows a breakdown of the goods' footprints in 2014.

Carbon Footprint

Data on the environmental impact of items were derived from existing LCAs (Life Cycle Assessments) of similar products (LCA-WG 2014; McNamara 2013; Stutz 2010, 2013; Dell 2014; Pihkola et al. 2010; Muthu 2015" with "Dell 2014; LCA-WG 2014; McNamara 2013; Muthu 2015; Pihkola et al. 2010; Stutz 2010; Stutz 2013; Tinsley 2006). LCAs include the impact generated over the whole life cycle of a product, process or service throughout its entire life cycle by quantifying material and energy inputs (consumption)

Table 2.3 Items possession and acquisition

Item	Total amount (2016)	Amount per capita (2016)	Acquisition rate* per year (average)	Acquisition rate* per capita (average)	Average age [yrs.]
cars	34	0.26	1/2	0.004	12
fridges	32	0.25	2	0.015	7
freezers	10	0.08	1/3	0.002	13
dishwashers	5	0.04	1/5	0.002	8
washing machines	9	0.07	1/5	0.002	10
laptops	97	0.75	4	0.032	5
PCs	7	0.05	1	0.002	4
printers	39	0.30	2	0.018	5
TVs	12	0.09	1	0.009	4

*new items only

Table 2.4 Goods' footprints

Goods	New items per person	CF [$kgCO_{2eq}$/person]	EF [gha/person]			
			energy land	cropland	forest land	total
office equipment and home appliances	0.10	17.72	0.005	0.000	0.000	0.005
vehicles	0.01	2.10	0.001	0.000	0.000	0.001
books, magazines and newspapers	95.00	24.99	0.006	0.000	0.031	0.037
garments	8.00	48.00	0.012	0.012	0.000	0.025
total		92.80	0.024	0.012	0.031	0.067

and outputs (emissions) during the extraction of raw materials, transport, production, distribution, use and disposal. In order to avoid double counting, we only accounted for the production phase and (when available) the transportation phase; in fact, the impact of the use phase for electrical items is already accounted as electricity consumption, while the end of life phase is accounted in the waste section, if happening during the observation year.

The goods CF is 92.80 $kgCO_{2eq}$/person. This includes both durable and non-durable items. Non-durable goods are the greatest contributors to the footprint (79%). This is due to the large number of consumables purchased in one year (see Table 2.4). Vehicle footprint is very small as a consequence of the low number of new vehicles purchased (see Table 2.3): most cars are second hand, and the number of cars per person is lower than 0.3 (the German average is 0.572 (Nationmaster)).

Ecological footprint

Goods EF is 0.067 gha/person. Almost 65% of the overall EF is due to the land (cropland and forest) needed to grow the raw material (cotton and cellulose) used to produce non-durable goods such as books, magazines and clothes. The remaining share of footprint is represented by the forest needed to absorb CO_2 emitted during the production and transportation of goods – that is, their CF.

2.4.3 Waste

Waste production is monitored yearly by the ecovillage, which provided data on the main categories of recyclable waste as well as mixed waste produced in the ecovillage over the observation year. Glass waste was not accounted because of the lack of information, but in Sieben Linden it is common practice to reuse most glass containers. Waste production in the ecovillage was 168 m³ per person, or 101 kg per person (conversion on the basis of the estimated average weight by waste type), excluding glass and organic waste. This value is noticeably lower than German average, which was 625 kg/person in 2015 (Eurostat). Organic waste is treated in an on-site treatment plant, and the compost obtained is used as fertiliser in the forest; therefore, it is not accounted in the footprint calculation. Table 2.5 shows a breakdown of waste production and footprints.

Table 2.5 Waste footprints

	Weight [kg/person]	CF [kgCO$_{2eq}$/person]	EF [gha/person] energy land
paper (recycle plant)	51.14	28.59	0.007
glass (recycle plant)	0.00	0.00	0.000
plastic (recycle plant)	11.72	3.97	0.001
electronics (recycle plant)	0.74	0.34	0.000
metals (recycle plant)	1.26	1.11	0.000
mixed waste (mainly incinerator)	35.79	68.36	0.018
total	*100.65*	*102.38*	*0.026*

Carbon Footprint

Waste CF is 102.38 kgCO$_{2eq}$/person. This only includes emissions generated by treating waste. The impact of the plant itself (construction and maintenance) is not accounted as it is included in the services (see Chapter 2.4.7). Mixed waste, that is not recyclable, represents more than 65% of total waste footprint (data on mixed waste impact from Lenaghan (2016)). Paper, which constitutes the largest share by weight, contributes to CF by 25% only (data on recycled waste impact form Turner et al. (2015)).

Ecological footprint

Waste EF is 0.026 gha/person. This only includes energy land needed to absorb CO$_2$ emissions generated by treating waste (that is, the CF of waste).

2.4.4 Travel

Between June 2014 and May 2015, twenty-four residents (nineteen adults and five children or teenagers, that is, 20% of the community) took part in a program to monitor travel outside the village; furthermore, an estimation on travel attitudes of each member has been supplied by the community. Combining this information, the distances travelled per resident have been calculated. Information about airplane travel was given separately for all members. Table 2.6 shows a breakdown of kilometres per person travelled outside the ecovillage, by purpose and means of transport.

A separate survey regarding daily movement routines inside the ecovillage (see Chapter 1.7), which sampled eleven residents, showed that on average more than one km is walked every day per person, which makes about 293 km per person per year. Adding these would bring the total distance travelled

Table 2.6 Breakdown of travel by purpose and means of transport

Reason of travel	Kilometres travelled [km/person]							
	car	bus	train*	airplane	ship	bicycle	on foot	total
work and school	2,180	1,815	1,366	231	0	145	15	5,753
purchases	500	51	151	0	0	66	21	788
leisure	1,478	183	2,693	0	2	123	48	4,527
political and social activities	70	6	65	169	0	0	0	311
total	4,228	2,054	4,276	400	2	335	85	11,417

*includes tramway and underground

Table 2.7 Travel footprints

Means of transport	Kilometres travelled [km/person]	CF [kgCO$_{2eq}$/person]	EF [gha/person] energy land
car	4,228.01	750.09	0.192
bus	2,054.37	208.97	0.054
train*	4,275.53	208.86	0.054
airplane	400.47	58.78	0.015
ship	1.90	0.22	0.000
total	*10,968.39*	*1,226.92*	*0.315*

*includes tramway and underground

per year to 11,710 km, of which walking and bicycling are about 6%, public transport 54%, air travel 3%, and private motor transport 36% (see Table 2.6).

Table 2.7 shows a breakdown of the travel footprints.

Carbon Footprint

Travel CF is 1,226.92 kgCO$_{2eq}$/person. The emissions have been calculated with reference to the kgCO$_{2eq}$ emitted per person per kilometre with each motorised means of transport (DEFRA 2016); these values refer to an average occupation rate, except for private cars where specific data on occupation were obtained through the monitoring program. Travel footprint is dominated by car travel (61%); bus and train have the same impact but distances travelled by train are double those by bus (see Table 2.7). The nearest operating train station is 30 km away, so a typical train journey begins with a 30-km bus ride. Walking and cycling have not been included, as they entail no emissions. The footprint associated with infrastructure construction and maintenance is accounted in the service footprint (see Chapter 2.4.7).

Ecological footprint

Travel EF is 0.315 gha/person. This only includes the energy land needed to absorb CO$_2$ emissions generated by motor vehicles (that is, the CF of transport).

2.4.5 Food

Data on food consumption were provided by the ecovillage, based on the quantities handled by the food cooperative (which provides food for the community kitchen in the Regiohaus and for the residents' home consumption) and those sold by the local shop (here, one can buy special foods such

Table 2.8 Food footprints

Food type	Weight [kg/person]	CF [kgCO$_{2eq}$/person]	EF [gha/person]				
			energy land	cropland	grazing land	sea	total
vegetables, fruits	488.96	314.60	0.081	0.128	0.000	0.000	*0.208*
cereals, bread, pasta	92.78	98.54	0.025	0.105	0.000	0.000	*0.130*
oils, spices	20.09	28.70	0.007	0.070	0.000	0.000	*0.077*
beverages	152.20	118.13	0.030	0.039	0.000	0.000	*0.069*
meat, fish, dairy products	92.26	137.54	0.035	0.000	0.110	0.004	*0.149*
sweets and other	76.27	26.08	0.007	0.081	0.000	0.000	*0.088*
total	*922.56*	*723.59*	*0.186*	*0.422*	*0.110*	*0.004*	*0.722*

as meat and chocolate). Data on the products' origin were also provided; the cooperative also produces organic vegetables and fruits on Sieben Linden land.

Most residents are vegetarian or vegan. Meat consumption in Sieben Linden is 97% lower than German average; while dairy products consumption is 10% lower (Heuer et al. 2015). However, the amount of food consumed by Sieben Linden residents – even when they are in the ecovillage – is somewhat underestimated (and therefore the food footprints) as some residents happen to buy some of their foodstuff outside the village, or occasionally have pizza delivered to their homes.

Table 2.8 shows the main categories that contribute to the food footprints.

Carbon Footprint

The food CF is 723.59 kgCO$_{2eq}$/person. This includes both production and transport of food (emissions per kilogram and kilometre calculated with EcoTransit online software (EcoTransit)). Transport accounts for less than 5% due to the residents' preference for local products (74% of total food purchase is from Germany and less than 5% is non-European). Food CF has been calculated with reference to the embodied energy of food products (data obtained from the Global Footprint Network); emissions have been obtained applying a world average primary energy carbon intensity (Sims 2007:261). Emissions generated by food production refer to conventional agriculture; as the ecovillage almost exclusively consumes organic products, this might imply some overestimation. On the other hand, organically self-produced vegetables and fruits have been assessed as zero impact (which might imply some underestimation).

Ecological footprint

The food EF is 0.722 gha/person. The forest area needed to absorb the CO_2 emitted during production and transportation of foodstuffs (that is, the CF) represents 26%; the remaining 74% corresponds to the biologically productive areas needed to grow primary products; grazing land footprint was calculated with reference to the average national grazing land value (GFN (b)), weighted by the weight of the products consumed (see Chapter 2.4.5: 3% of meat consumption + 90% of the consumption of dairy products). Vegetables and cereals are the largest contributors to the food EF. This is mainly related to the area required to grow them. However, vegetables and cereals are characterised by a lower unitary footprint than dairy products; indeed, vegetables account for more than 50% of total consumption by weight, whereas their footprint is less than 30% of total footprint. Meat and dairy products are little consumed (10%) but imply a high footprint (20%).

2.4.6 Built-up land

Built-up land reflects the bioproductive area compromised by anthropic infrastructure. Built-up land information for the ecovillage has been extracted from plans and technical drawings. This includes the area occupied by residential and communal buildings, trailers and roads. Also because of the community's lifestyle which privileges the use of collective space (see Chapter 1.7), the buildings do not take much land. Gravel paths through the ecovillage have not been accounted as built-up land (which equals land which has lost its biocapacity) since they could be easily reconverted into green areas; however, they have neither been accounted as productive surface in the biocapacity account (see Chapter 5.2).

Built-up land EF is 0.034 gha/person. Table 2.9 shows built-up land actual extension (in ha) compared to Ecological Footprint (in gha).

2.4.7 Services

Data on services cannot be collected at the local level; national average data need to be used instead. According to EUREAPA,[4] the contribution of services to the CF for German citizens is 3.950 $kgCO_{2eq}$, while their EF is 1.32 gha/person. This includes government services (e.g. health, education) and capital investment (e.g. infrastructure).

Table 2.9 Built-up land footprint

	Built-up land [ha/person]	EF [gha/person]
buildings	0.004	0.009
trailers	0.001	0.002
roads	0.009	0.022
total	*0.013*	*0.034*

2.5 Overall footprints and comparisons

2.5.1 Carbon Footprint

The total CF of Sieben Linden is 810,189.90 $kgCO_{2eq}$, that is 6,379.45 $kgCO_{2eq}$/person. Table 2.10 and Figure 2.1 show a breakdown of the categories.

The CF is dominated by the impact of services, which represent 62% of the total footprint. This part is ascribed to Sieben Linden residents as German citizens, and cannot be directly influenced by their lifestyle choices.

Excluding the services footprint, the residents' footprint is 2,429.45 $kgCO_{2eq}$/person. The greatest contributor to the locally controlled share of the footprint is travel (50%), mainly due to the use of cars (which represent 61% of it). Food footprint is 30%, mostly due to the production phase (transportation is just 5%). Energy footprint is less than 12%, and it is mainly caused by firewood combustion. Waste and goods do not much affect the overall footprint, as their sum represents about 8% of the locally controlled share of CF.

Table 2.10 Overall Carbon Footprint

Components		CF [kgCO_{2eq}/person]	CF [%]
energy	electricity	0.00	0.00
	heating	216.70	3.40
	propane gas	67.05	1.05
goods	items	92.80	1.45
waste	waste	102.38	1.60
travel	travel	1,226.92	19.24
food	food	723.59	11.34
other	services	3,950.00	61.91
total		*6,379.45*	*100.00*

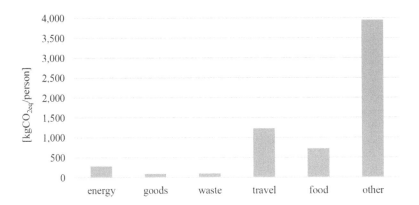

Figure 2.1 Overall Carbon Footprint

2.5.2 Carbon Footprint comparisons

Sieben Linden vs. German average

Table 2.11 and Figure 2.2 show the CF of Sieben Linden residents compared to that of average German citizens. Data on German CF (as of 2014) are available through DESTATIS (DESTATIS). These data have been subdivided according to EUREAPA categories (Eureapa), as DESTATIS data are arranged according to categories that are spurious with those adopted in the present study.

As the national data do not include a heading for waste (possibly because it is associated with the category of products that generate it), for the sake of comparison the impact of waste produced by ecovillage residents has been added to that of goods. Values for services are by definition the same, as we employed the national average value for Sieben Linden as well; whereas we were not able to assess the impact of "other services" (which include communication, leisure, tourist facilities, etc.) for Sieben Linden.

In all categories (services excluded), Sieben Linden fares significantly better than the German average. Energy and goods are almost 90% lower, while food is 36% lower. Travel is approximately one half of the national mean. In sum, the overall CF of Sieben Linden is 48% of the German average (27% if the services' CF is excluded).

Sieben Linden 2014–2002

A group from the University of Kassel calculated the carbon and energy footprints of Sieben Linden for 2002, when the residents totalled fifty-two

Table 2.11 Comparison between Sieben Linden and German average carbon footprints

Category	Germany (2014)		Sieben Linden (2014)		Ratio (Sieben Linden/ Germany* 100)
	CF [kgCO$_{2eq}$/ person]	CF [%]	CF [kgCO$_{2eq}$/ person]	CF [%]	
energy	2,533.34	19.25	283.75	4.45	11.20
goods and waste	1,989.89	15.12	195.18	3.06	9.81
travel	2,374.49	18.04	1,226.92	19.24	51.69
food	1,998.25	15.18	723.59	11.34	36.21
services	3,950.00	30.01	3,950.00	61.91	100.00
other services*	314.05	2.38	–	–	–
total	*13,160.00*	*100.00*	*6,379.45*	*100.00*	*48.48*

*include communication, leisure, tourist facilities, etc.

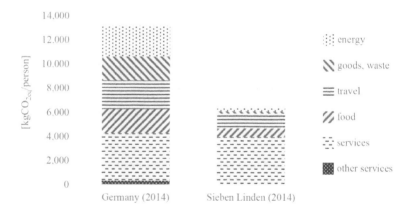

Figure 2.2 Comparison between Sieben Linden and German average carbon footprints

(Dangelmeyer et al. 2004). Methodology, data retrieval and boundaries differ from our own study, making the results only partially comparable. However, an interesting overview on consumption patterns for 2002 could be extracted.

Table 2.12 and Figure 2.3 show a comparison between consumption in the ecovillage in 2002 and 2014.

Table 2.12 Comparison of consumption by category, 2014–2002

	Electricity [kWh/p*y]		Firewood [rm/p*y]	Propane gas [kg/p*y]	Water [l/p*y]	Food [kg/p*y]	Travel [km/p*y]
2002	317.0		2.5	20.6	83.6	720	12,944
2014	148.5*	442.1**	1.5	18.1	61.8	690	11,417
difference (2014–2002) %	−53.2	+39.5	−40.0	−12.1	−26.1	−4.2	−11.8

*excluding consumption covered by self-production from PV panels
**total electricity consumption

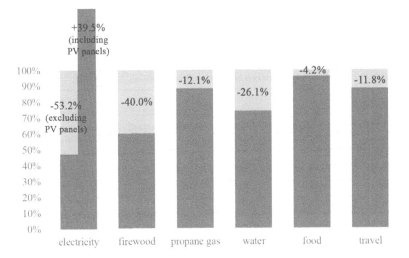

Figure 2.3 Comparison of consumption by category, 2014–2002

It may be interesting to remark that in all areas there has been a reduction; the only exception is electricity. However, the increase is more than effaced by self-production. Yet, it must be said that some values as of 2014 seem to be more virtuous than an average yearly value calculated on a longer time span. In particular, the average value of firewood consumption (2009–2015) has been 2.3 rm/p*y; this value obviously fluctuates depending on the mean temperature of each winter. During the very last winter, though, a decrease in temperature was not coupled to an increase of firewood consumption. It is too early to draw conclusions, but this might be the effect of the ever-increasing ratio of superinsulated buildings to trailers (see also Chapter 4.2).

A comparison with the 2002 carbon footprint can only be done by taking into account two aspects:

- data concerning 2002 only account for residents' impact within the eco-village; on the other hand, data from 2014 were normalised to represent the whole year impact (see Chapter 2.3). Therefore, in this comparison non-normalised data were used;
- because of the lack of data in 2002, some categories (solar panels, goods, waste, services) have been excluded from the comparison.

Table 2.13 and Figure 2.4 show a comparison between CO_2 emissions in 2002 and 2014.

In all categories, Sieben Linden reduced its Carbon Footprint since 2002. Food footprint is 36% lower, possibly also thanks to the higher number of residents sharing the facilities. Travel and energy are both 19% lower. The overall CF of Sieben Linden in 2014 is 25% lower than in 2002.

Table 2.13 Comparison between Sieben Linden CF in 2002 and 2014

	2002		2014		Ratio (2014 / 2002* 100)
	[kgCO_2eq/person]	[%]	[kgCO_2eq/person]	[%]	
energy	238.00	9.06	193.93	9.82	81.48
travel	1,530.00	58.22	1,234.00	62.51	80.65
food	860.00	32.72	546.19	27.67	63.51
total	2,628.00	100.00	1,974.12	100.00	75.12

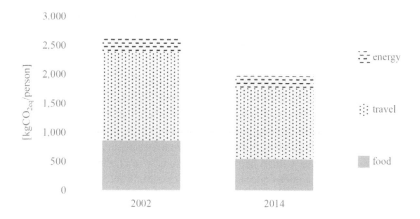

Figure 2.4 Comparison between Sieben Linden CF in 2002 and 2014

2.5.3 Ecological Footprint

The total EF of an average Sieben Linden resident is 3.06 gha; the footprint of the whole ecovillage is 391.67 gha. Table 2.14 and Figure 2.5 show a breakdown of the categories contributing to the footprint.

Table 2.14 Overall Ecological Footprint

Components		EF [gha/person]							%
		energy land	cropland	grazing land	forest land	sea	built-up land	total	
energy	electricity	0.000	0.000	0.000	0.000	0.000	0.000	*0.000*	0.00
	heating	0.006	0.000	0.000	0.557	0.000	0.000	*0.564*	18.39
	propane gas	0.017	0.000	0.000	0.000	0.000	0.000	*0.017*	0.56
goods	items	0.024	0.012	0.000	0.031	0.000	0.000	*0.067*	2.19
waste	waste	0.026	0.000	0.000	0.000	0.000	0.000	*0.026*	0.86
travel	travel	0.315	0.000	0.000	0.000	0.000	0.000	*0.315*	10.27
food	food	0.186	0.422	0.110	0.000	0.004	0.000	*0.722*	23.55
other	built-up land	0.000	0.000	0.000	0.000	0.000	0.034	*0.034*	1.11
	services	–	–	–	–	–	–	*1.320*	43.07
total		*0.574*	*0.434*	*0.110*	*0.588*	*0.004*	*0.034*	*3.064*	*100.00*

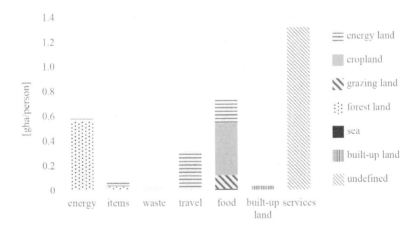

Figure 2.5 Overall Ecological Footprint

The EF is dominated by the impact of services, which represents 43% of total footprint. This part is ascribed to Sieben Linden residents as German citizens and cannot be directly influenced by their lifestyle choices. Excluding services, the residents' footprint is 1.74 gha/person. The greatest contributor to the locally controlled part of the footprint is food (41%), mainly because of the large extent of cropland needed to grow vegetables and cereals. Energy footprint is 33%, mostly forest for firewood production. Transport footprint is less than 20% and is mainly caused by the use of cars. Waste, goods and built-up land footprint don't affect much the overall footprint and together represent less than 7% of total footprint.

2.5.4 Ecological Footprint comparisons

Sieben Linden vs. German average

Table 2.15 and Figure 2.6 show the ecological footprint of a Sieben Linden resident compared to that of the average German citizen. Data on the German footprint are available through the Global Footprint Network open data platform (GFN (b)); these data have been subdivided according to EUREAPA categories (Eureapa).

As the national average values do not include a heading for waste – possibly because it is associated with the category of products that generate it – the impact of waste produced by ecovillage residents has been summed to that of goods. Values for services are by definition the same as we applied

Table 2.15 Comparison between Sieben Linden footprint and average German footprint

Category	Germany (2013)		Sieben Linden (2014)	
	EF [gha/person]	%	EF [gha/person]	%
energy	0.684	12.5	0.581	18.9
goods and waste	0.886	16.2	0.093	3.0
travel	0.658	12.0	0.315	10.3
food	1.413	25.8	0.722	23.5
built-up land	0.149	2.7	0.034	1.1
services	1.320	24.1	1.320	43.0
other services*	0.360	6.6	–	–
total	*5.470*	*100.0*	*3.064*	*100.0*

*includes communication, leisure, tourist facilities, etc.

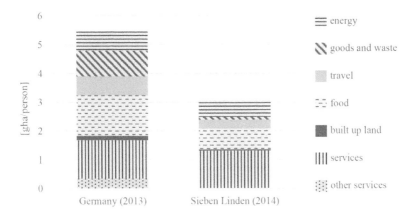

Figure 2.6 Comparison between Sieben Linden footprint and German average
footprint

the national average for Sieben Linden as well; whereas we were not able
to assess the impact of "other services" (which includes communication,
leisure, tourist facilities, etc.) for Sieben Linden, and we left it blank.

The ecovillage's impact of energy is slightly lower than the German
average – yet it is relevant to underline that most of such impact is due
to the exploitation of a renewable resource (firewood), while that of Germany is mainly due to CO_2 emissions from the burning of non-renewable
fuels. In all the remainder categories Sieben Linden fares significantly
better than the average. Goods and waste are almost 90% lower, probably
thanks to the frugal buying patterns coupled with sharing and exchanging
second-hand items. In both travel and food categories the ecovillage's
EF is approximately one-half of the national mean. In sum, the overall
EF of Sieben Linden is 56% of the German average (46% if services are
excluded).

Sieben Linden vs. fair share Ecological Footprint

The Ecological Footprint of every individual and country can be expressed
as the number of Earths ("Planet Equivalent") it would take to support that
footprint if everyone lived like that individual, or the average citizen of that
country; it is the ratio of the EF to the per capita biological capacity available on Earth (GFN (b)). Such capacity was 1.7 gha in 2013 (GFN (c)). The
Vales (2013) defined this as the "fair share" ecological footprint.

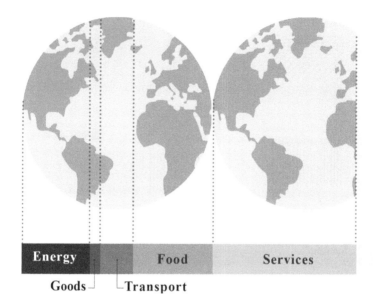

Figure 2.7 Planet equivalents for Sieben Linden

Excluding services footprint, we calculated Sieben Linden residents' footprint at 1.74 gha/person in 2014 (see Table 2.14). This would almost equal the "fair share" EF. However, adding the impact of services, the overall footprint grows to 3.06 gha/person, which means 1.8 Planet Equivalents (see Figure 2.7). In 2013, the world average Ecological Footprint of 2.87 gha equalled 1.7 "planets."

Sieben Linden vs. other ecovillages

The results obtained have been compared against six selected cases:

• BedZED

(Hodge and Haltrech 2009);

• the Findhorn Foundation and Community

(Tinsley and George 2006);

• Steward Community Woodland

(Knight 2008);

• Toarp ecovillage

(Haraldsson and Svensson 2000);

*   Krishna Valley

(Lánczi 2009);

*   OUR Ecovillage

(Giratalla 2010).

These cases were chosen according to two criteria:

*   affinity: selected cases show environmental and social similarities with Sieben Linden and are located in contexts that are comparable with it, both climatically and socio-economically (Europe, UK, Canada). Moreover, all communities are based on sustainability principles, which affect the ways of everyday life;
*   availability: clear explanation of methodology, completeness of survey.

The methodological consistency has been checked against the fourteen compatibility criteria identified by Lewan and Simmons (2001). The main discrepancies include the lack of data for "goods" (two cases), waste (four cases) and built-up land (five cases).

We decided to keep the impact due to public services out of the comparison as they depend from superordinate choices which are out of the residents' control.

Table 2.16 and Figure 2.8 show a comparison between Sieben Linden's EF and other communities' EF.

The EF of all cases is far below the German average. Two communities have a "local" EF lower than the "fair share" EF, that is to say within the world mean biocapacity – however, the addition of the EF due to public

Table 2.16 Comparison between Sieben Linden's EF and other communities' EF

	Country	EF (gha/person)						
		home & energy	goods	waste	travel	food	built-up land	total
Sieben Linden	DE	0.58	0.07	0.03	0.31	0.72	0.03	*1.74*
Findhorn Foundation	UK	0.29	0.30	0.20	0.37	0.42	n/a	*1.58*
BedZED	UK	0.77	0.79	n/a	0.75	1.22	n/a	*3.53*
Steward Community Woodland	UK	0.24	0.64	n/a	0.30	0.66	n/a	*1.84*
Toarp	SE	1.15	0.15	0.04	0.42	0.93	n/a	*2.69*
OUR ecovillage	CA	0.28	n/a	0.30	0.67	1.12	0.15	*2.52*
Krishna Valley	HU	0.29	n/a	n/a	0.18	0.42	n/a	*0.89*

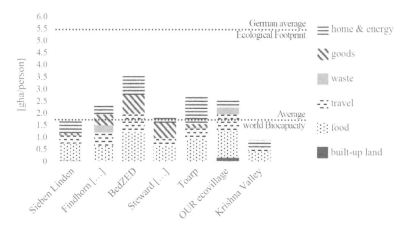

Figure 2.8 Comparison between Sieben Linden's EF and other communities' EF

administration and other services would inevitably make them pass such threshold. Moreover, some of the "local" impact categories were also not included in these studies (see Table 2.17); one ought to recall this to avoid unfair comparisons. For instance, values of Krishna Valley are particularly low also because they do not include several such categories. This said, in spite of covering all impact categories, Sieben Linden's EF is lower than the average value of the six reference cases.

Notes

1 The Carbon Trust is an independent, expert partner of leading organisations around the world, helping them contribute to and benefit from a more sustainable future through carbon reduction, resource efficiency strategies and commercialising low carbon technologies (Carbon Trust (b)).

2 The Global Footprint Network is an international non-profit organisation founded in 2003 to enable a sustainable future where all people have the opportunity to thrive within the means of one planet (GFN (d)).

3 Data on firewood consumption were provided in raummeter [rm] (that is, a 1 m³ pile of stacked woodpile with air spaces) and converted in kWh with reference to a net calorific value for firewood of 14,4 MJ/kg (Francescato 2004:16); data on propane gas consumption were provided in kg and converted in kWh with reference to a net calorific value for propane of 46,35 MJ/kg (Guadagni 2010:243).

4 EUREAPA is an online scenario modelling and policy assessment tool created for the One Planet Economy Network. It uses a sophisticated economic input-output model to understand the environmental pressures associated with consumption activities. It contains baseline data on the economy, greenhouse gas emissions, ecological footprints and water footprints for every EU member state and 16 other countries and regions of the world (Eureapa).

References

Carbon Trust (a), *Carbon Footprinting Guide* [online]. Available from: www.carbontrust.com/media/44869/j7912_ctv043_carbon_footprinting_aw_interactive. pdf [last viewed Sept. 2018].

Carbon Trust (b), *About Us* [online]. Available from: www.carbontrust.com/aboutus/ [last viewed Sept. 2018].

Dangelmeyer, Peter et al. (eds.), *Ergebnisse des Vorhabens Gemeinschaftliche Lebens- und Wirtschaftsweisen und ihre Umweltrelevanz*, Kassel: Wissenschaftliches Zentrum für Umweltsystemforschung – Universität Kassel, 2004.

Defra (Department of Energy & Climate Change and Department of Environment Food & Rural Affairs), *UK Government GHG Conversion Factors for Company Reporting 2016*, 2016. Available from: www.gov.uk/government/publications/ greenhouse-gas-reporting-conversion-factors-2016 [last viewed Sept. 2018].

Dell, *Product Environmental Aspects Declaration for Dell Colour Printer – C2660dn* [online], 2014. Available from: https://i.dell.com/sites/doccontent/corporate/corpcomm/en/Documents/Dell-C2660dn-AD-14-E534.pdf [last viewed Sept. 2018].

DESTATIS, *Environmenta Economic Accounting. Material and Energy Flows* [online]. Available from: www.destatis.de/EN/FactsFigures/NationalEconomy Environment/Environment/EnvironmentalEconomicAccounting/Material EnergyFlows/Tables/ProductionFactorsPollutants.html [last viewed Sept. 2018].

EcoTransit, *EcoTransit World* [online]. Available from: www.ecotransit.org/ [last viewed Sept. 2018].

EIA (Energy Information Administration U.S. Department of Energy), *Instructions for Form EIA-1605. Voluntary Reporting of Greenhouse Gases. Revised Pursuant to 10 CFR Part 300 Guidelines for Voluntary Greenhouse Gas Reporting*, 2010. Available from: www.reginfo.gov/public/do/DownloadDocument?objec tID=2653501 [last viewed Sept. 2018].

Eureapa, *One Planet Economy Network* [online]. Available from: www.eureapa. net/explore/?impactgroup_id=1&region_id=9&footprintgroup_id=0 [last viewed Sept. 2018].

Eurostat, *Municipal Waste Statistics* [online]. Available from: http://ec.europa.eu/ eurostat/statistics-explained/index.php/Municipal_waste_statistics [last viewed Sept. 2018].

Francescato, Valter; Eliseo Antonini; Giustino Mezzalira, *L'energia del legno: nozioni, concetti e numeri di base*, Torino: Regione Piemonte, 2004. Available from: www.regione.piemonte.it/foreste/images/files/pubblicazioni/energia_legno.pdf [last viewed Sept. 2018].

Galli, Alessandro et al., "An exploration of the mathematics behind the ecological footprint", *International Journal of Ecodynamics*, 2, 4, 2007, pp. 250–257. doi:10.2495/ECO-V2-N4-250-257

GFN (a), *Global Footprint Network Glossary* [online]. Available from: www. footprintnetwork.org/resources/glossary/ [last viewed Sept. 2018].

GFN (b), *Ecological Footprint Analyze by Land Type* [online]. Available from: http:// data.footprintnetwork.org/#/analyzeTrends?type=EFCtot&cn=79 [last viewed Sept. 2018].

GFN (c), *Ecological Wealth of Nations* [online]. Available from www.footprint-network.org/content/documents/ecological_footprint_nations/biocapacity_per_capita.html [last viewed Sept. 2018].

GFN (d), *About us* [online]. Available from: www.footprintnetwork.org/about-us/ [last viewed Sept. 2018].

Giratalla, Waleed, *Assessing the Environmental Practices and Impacts of Intentional Communities: An Ecological Footprint Comparison of an Ecovillage and Cohousing Community in Southwestern*, Vancouver: University of British Columbia, 2010. Available from: https://pics.uvic.ca/sites/default/files/uploads/publications/Giratalla_Thesis.pdf [last viewed Sept. 2018].

Guadagni, Andrea (ed.), *Prontuario dell'ingegnere*, Milano: Hoepli, 2010.

Haraldsson, Hördur; Mats G.E. Svensson, *Is Ecological Living Sustainable? A Case Study From Two Swedish Villages in South Sweden*, Lund: Lund University Centre for Applied System Dynamics, 2000. Available from: www.systemdynamics.org/assets/conferences/2000/PDFs/haraldss.pdf [last viewed Sept. 2018].

Heuer, Thorsten et al., "Food consumption of adults in Germany: Results of the German National Nutrition Survey II based on diet history interviews", *The British Journal of Nutrition*, 113, 10, 2015, pp. 1603–1614. doi:10.1017/S0007114515000744

Hodge, Jessica; Julia Haltrech, *BedZed Seven Years on the Impact of the UK's Best Known Eco-village and Its Residents*, Wallington: BioRegional, 2009. Available from: www.bioregional.com/wp-content/uploads/2014/10/BedZED_seven_years_on.pdf [last viewed Sept. 2018].

Kitzes, Justin, *Ecological Footprint Standards 2009*, Oakland: Global Footprint Network, 2009. Available from: www.footprintnetwork.org/content/images/uploads/Ecological_Footprint_Standards_2009.pdf [last viewed Sept. 2018].

Knight, William, *Ecological Footprint Report of Steward Community Woodland*, Looe, UK: 4th World Ecological Design, 2008. Available from: www.stewardwood.org/pdf/03_ecological_footprint.pdf [last viewed Sept. 2018].

Lánczi, Dániel Csaba, *Practice of Sustainability in an Eco Village: Ecological Footprint of KrishnaValley in Hungary*, Budapest: Eötvös Lóránd University, Faculty of Science, Department of Environment and Land Geography, 2009. Available from: https://gen-europe.org/uploads/media/Eco_Footprint_Krishna_Valley.pdf [last viewed Sept. 2018].

LCA-WG (Life Cycle Assessment – Working Group) of the Environmental Technical Expert Committee of the Japan Electrical Manufacturers' Association, *Report on Life Cycle Inventory (LCI) Analyses of Refrigerators*, 2014. Available from: www.jema-net.or.jp/English/businessfields/environment/data/report_lci.pdf [last viewed Sept. 2018].

Lenaghan, Michael, *The Scottish Carbon Metric: A National Carbon Indicator for Waste: 2013 Update to the Technical Report*, Stirling, Scotland: Zero Waste Scotland, 2016. Available from: www.zerowastescotland.org.uk/sites/default/files/2013%20Carbon%20Metric%20-%20Technical%20Report.pdf [last viewed Sept. 2018].

Lewan, Lillemor; Craig Simmons, *The Use of Ecological Footprint and Biocapacity Analyses as Sustainability Indicators for Sub-national Geographical Areas:*

A Recommended Way Forward. Final report, Prepared for Ambiente Italia ECPI (European Common Indicators Project), 2001. Available from: http://manifestinfo. net/susdev/01EUfootprint.pdf [last viewed Sept. 2018].

Lin, David et al., *Working Guidebook to the National Footprint Accounts: 2016 Edition*, Oakland: Global Footprint Network, 2016. Available from: www.footprintnetwork.org/content/documents/National_Footprint_Accounts_2016_Guidebook. pdf [last viewed Sept. 2018].

McNamara, David, *Life Cycle Assessment of Washing Machine* [online]. Updated December 2013. Available from: https://mcnamaradavid.files.wordpress.com/2013/ 12/et4407_lca_washingmachine1.pdf [last viewed Sept. 2018].

Menzies, Gillian; Ya Roderick, "Energy and carbon impact analysis of a solar thermal collector system", *International Journal of Sustainable Engineering*, 3, 1, 2010, pp. 9–16. doi:10.1080/19397030903362869

Nationmaster, *Motor Vehicles Per 1000 People: Countries Compared* [online]. Available from www.nationmaster.com/country-info/stats/Transport/Road/Motor-vehicles-per-1000-people [last viewed Sept. 2018].

Pihkola, Hanna et al., *Carbon Footprint and Environmental Impacts of Print Products From Cradle to Grave. Results From the LEADER Project (Part 1)*, Vuorimiehentie, Finland: VTT Technical Research Centre of Finland, 2010. Available from: www.vtt.fi/inf/pdf/tiedotteet/2010/T2560.pdf [last viewed Sept. 2018].

Sims, Ralph E.H.; Robert N. Schock (eds.), "Energy supply", in B. Metz et al. (eds.), *Climate Change 2007: Mitigation. Contribution of Working Group III to the Fourth Assessment Report of the Intergovernmental Panel on Climate Change*, Cambridge: Cambridge University Press, 2007. Available from: www.ipcc.ch/ pdf/assessment-report/ar4/wg3/ar4-wg3-chapter4.pdf [last viewed Sept. 2018].

Stutz, Markus, *Carbon Footprint of a Typical Business Laptop From Dell*, 2010 [online]. Available from: https://i.dell.com/sites/content/corporate/corp-comm/en/ Documents/dell-laptop-carbon-footprint-whitepaper.pdf [last viewed Sept. 2018].

Stutz, Markus, *Carbon Footprint of a Typical 19" Business Monitor From Dell*, 2013 [online]. Available from: https://i.dell.com/sites/doccontent/corporate/corp-comm/en/Documents/display-white-paper.pdf [last viewed Sept. 2018].

Subramanian Senthilkannan Muthu (ed.), *Handbook of Life Cycle Assessment (LCA) of Textiles and Clothing*, Cambridge: Woodhead Publishing, 2015.

Sullivan, John Lorenzo; Andrew Burnham; Michael Wang, *Energy Consumption and Carbon Emission Analysis of Vehicle and Component Manufacturing*, Lemont: Argonne National Laboratory, 2010. Available from: https://publications.anl. gov/anlpubs/2010/10/68288.pdf [last viewed Sept. 2018].

Tinsley, Stephen; Heater George, *Ecological Footprint of the Findhorn Foundation and Community*, Forres, Scotland: Sustainable Development Research Centre, 2006. Available from: ttps://www.ecovillagefindhorn.com/docs/FF%20Footprint. pdf [last viewed Sept. 2018].

Turner, David A.; Ian D. Williams; Simon Kemp, "Greenhouse gas emission factors for recycling of source-segregated waste materials", *Resources, Conservation and Recycling*, 105, Part A, 2015, pp. 186–197. doi:10.1016/j.resconrec.2015.10.026

Vale, Robert; Brenda Vale (eds.), *Living Within a Fair Share Ecological Footprint*, Abingdon: Routledge, 2013.

Wackernagel, Mathis; Nicky Chambers; Craig Simmons, *Sharing Nature's Interest: Ecological Footprints as an Indicator of Sustainability*, London: Earthscan Publications Ltd, 2000.

Wiedmann, Thomas; Jan Minx, "A definition of 'carbon footprint'", in C. C. Pertsova (eds.), *Ecological Economics Research Trends*, Hauppauge, NY: Nova Science Publishers, 2008, pp. 1–11.

3 The environmental impact of the Sieben Linden buildings

Martina Gerace

The categories of environmental impact considered in this study are:

- Embodied Energy (EE) or Primary Energy Intensity (PEI): it represents the amount of energy needed to produce, transport to the site and dispose of a product or material or to provide a service. In the case of buildings, Embodied Energy is usually measured as the amount of energy pertaining to the unit of building material, component or system. It can be expressed in megajoules per unit of weight (MJ/kg) or, less commonly, volume (MJ/m^3). The EE value of a product depends not only on the production process, but also on factors such as the energy efficiency of the machinery used, the distance from the raw materials supply, the modes of transport, the sources of energy, and the local energy mix.
- Global Warming Potential (GWP): represents the amount of greenhouse gas emissions in the atmosphere related to production, transport to the site of use and disposal of a product or material or to provide a service. It is measured in $kgCO_{2eq}$, i.e. kilograms of carbon dioxide (and other polluting gases made equivalent to carbon dioxide) released to supply a unit of product, and therefore it is generally referred to as Embodied Carbon (EC) or Carbon Footprint (CF).

3.1 Case studies

This section of the study deals with the analysis of two Sieben Linden buildings: Libelle and Villa Strohbunt, for which sufficient data were available.

Libelle is a residential building with a gross internal area (GIA) of 379 m² on two storeys, completed in 2012, designed by architect Dirk Scharmer.

The building features continuous foundations of reinforced concrete, a layer of gravel and a concrete screed. The loadbearing structure is made of wooden elements embedded in the perimeter walls, which are infilled with straw bales and plastered with lime externally and clay inside. The self-supporting internal partition walls are made of calcium silicate blocks 175 mm wide, plastered with 15 mm clay. The roof is single pitched sloping to the north; timber rafters are 6 × 36 cm and the space between them is infilled with straw insulation. "Green" building materials were locally sourced from the market and the house was erected by professionals.

Villa Strohbunt is a two-storey, 105 m² GIA building designed and self-built between 2001 and 2004 by Björn Meenen, Martin Stengel and Silke Hagmaier of the "Club99" with the help of hundreds of volunteers, attracted by the interest in straw bale building and manual labour, and the interaction with a radical community. The construction process was thoroughly experimental and based on the "resource diet" principle: in order to achieve 90% savings, they agreed to dispense with the use of machines, machine-made building materials and instead recycle materials as much as possible. There were only three exceptions: a quarter bag of cement for a few unavoidable joints in the foundation and the chimney; mechanical pressing of straw bales for walls, floors and ceilings; a few cubic metres of engineered timber panels in stairs and shelves (Hagmaier in Stanellé 2017:39).

It was created as a shared space but it is now used as a living space.

Its foundation of twenty-four granite reused plinths supports the ground floor beams above grade. It has a half-timber structure (*Fachwerk*) in round logs of pine, placed inside the perimeter walls made of clay-plastered straw bales. The internal space is free from columns thanks to the use of a solid timber floor (*Dippelbaumdecke*). The partition walls are made of clay bricks, usually plastered. The timber used in the construction comes from the ecovillage's forest, hand-felled and hand-worked, and transported by horse to the building site; it was assembled using traditional tools. Motor machinery was only employed to transport foundation stones, straw bales and clay to the site. Straw was partly from the first organic harvest of Sieben Linden fields and partly purchased in the area; clay and sand were dug from the ground of Sieben Linden. Other construction products such as foundation stones, pantiles, windows and gutters have been supplied locally; some of them were salvaged from other buildings (Wiegand et al. 2006).

3.2 Boundary of study and source of data

We chose to draw as much data as possible from a single source using transparent principles and methods. Thus reference was made to the Inventory of Carbon and Energy (ICE) of the University of Bath (Hammond 2011), which contains relevant information on data sources. The four criteria used for their selection were (Hammond 2008):

1 compliance with ISO 14040 and 14044;
2 definition of system boundaries from cradle to gate;
3 country of origin: for most of the materials it was necessary to refer to international sources but with a preference for UK sources;
4 year: the most recent ones have been preferred, in particular as regards the Embodied Carbon.

Data in the ICE do not include carbon absorbed and stored in plants (biosequestration). When a tree grows, it subtracts carbon dioxide from the atmosphere, which becomes part of its structure; this carbon remains bound in the material until it is burnt or decomposes. To widen our analysis, we decided to perform calculations both including and excluding carbon stored in vegetal materials; data missing from the ICE were obtained from other sources, such as Berge (2009). The boundaries adopted in this study are the same of the ICE: from cradle to gate, i.e. from the extraction phase to manufacturing phase. The energy used in the construction phase is not accounted for as not enough data was available. The energy consumption in the use phase is included in Chapter 2.

The main source of information related to Libelle was the bill of quantities, integrated by technical drawings and building photos that allowed to resolve some inconsistencies or difficulties in identifying the elements. (In practice, a complete set of new technical drawings has been created to give every building part its physical place and cross-check the quantities employed.) Items generically listed in the bill of quantities (that is, not associated with a specific product) were associated with building products available on the German market. As metal joints were not present in the bill of quantities, possibly because they were left to the carpenter's judgement, they were counted as 1 kg/m^2 gross floor area. It was not possible to ascertain to which elements 17% of the total timber appearing in the bill of quantities refers – however, such amount was included in the calculation.

Regarding Villa Strohbunt, a synthetic bill of quantities contained in a study by the Institute for Energy Technologies of TU Berlin (2006) was used. All salvaged elements – such as foundation stones, floor tiles, wooden ties connecting the straw bales, windows, doors, finishing plaster layer, roof tiles – have not been accounted. This approach is shared by the aforementioned study, since the EE and EC of elements ought only to be attributed to the building where they were used for the first time. It was not possible to trace the exact amount of steel in the building but it can be rounded down to zero because the timber joints interlock without mechanical fasteners and the few nails used in the building were salvaged.

In this part of the analysis the technical services are not taken into account, as information was available only for Libelle. This choice was made so to facilitate results comparability, as LCA studies of buildings usually do not include the technical services. However, the technical services of Libelle are analysed in Chapter 3.4.

3.3　PEI and GWP calculations

In order to calculate the PEI, it was necessary to associate one or more material(s) to each item listed in the bill of quantities and calculate its total weight, when not specified. Each material was then associated with the corresponding value of Embodied Energy [MJ/kg] taken from the ICE. Data missing were obtained from other sources: for cellulose, clay and lightweight concrete Berge (2009); for silicone, calcium silicate and titanium zinc Hegger et al. (2008); for jute van Dam (2004:6); for gypsum fibre board Fermacell data sheets; for reed wattle FNR (2014:23); for hemp FNR (2008:78).

In the case of Villa Strohbunt, clay, as well as pine and fir wood were obtained from the adjacent forest with no machinery; timber was dried naturally. For this reason, their EE was approximated to zero. Table 3.1 and Figure 3.1 show the results obtained.

The total weight of Libelle is 391.14 tonnes (i.e. 1.03 t/m^2 GIA). As 10 people live in the building, this equals 39.11 t/person.

Global PEI for Libelle is 1,308 GJ, which translates into 3.45 GJ/m^2 GIA, and 130.71 GJ/person.

The global weight of Villa Strohbunt is 104 tonnes (0.99 t/m^2 GIA), but salvaged material constitutes the 23% of the building by weight, so the weight of the materials taken into account in the subsequent calculations (PEI, GWP) is 80.45 tonnes. At the present six-persons occupation, such weight is 17.33 t per capita.

Global PEI for Villa Strohbunt is 82.72 GJ, which results in 0.79 GJ/m^2 GIA, and 13.79 GJ/person.

To calculate the GWP, it was decided to not use the ICE database coefficients directly – instead, values were normalised to the mix of German energy sources for the major industries, as by the Federal Statistical Office of Germany (DESTATIS 2015). Regarding conversion factors from energy to CO_{2eq} emissions entailed by fossil fuels, data from the German Federal Environment Agency (UBA) were used (Juhrich 2016:45–47). The German electricity mix was obtained from DESTATIS and refers to 2016 (Kono et al. 2017:3).

The results obtained are approximations: first because fossil fuels conversion factors return the $kgCO_2$ only and not total amount of GHGs; secondly because only emissions due to burning fuels were considered, not the chemical reactions that occur during the processing of certain materials (such as cement calcination).

Table 3.1 Libelle mass, PEI and GWP and Villa Strohbunt mass, PEI and GWP

Material	Libelle			Villa Strohbunt		
	weight [t]	PEI [MJ]	GWP [kgCO$_{2eq}$]	weight [t]	PEI [MJ]	GWP [kgCO$_{2eq}$]
steel	1.87	38,812	3,852	–	–	–
titanium zinc, aluminium	1.07	141,742	14,068	–	–	–
polyethylene, polypropylene	0.83	69,682	8,713	–	–	–
epoxy resin	0.34	47,154	5,896	–	–	–
paint	0.82	48,624	4,336	–	–	–
fibreglass	0.13	3,610	319	–	–	–
glass	2.84	42,594	3,762	0.54	–	–
granite	–	–	–	17.96	–	–
gravel, sand	73.89	6,133	542	33.46	2,710	239
concrete, cement	88.08	67,582	5,964	0.02	84	7
lime	11.11	58,863	5,199	–	–	–
calcium silicate	97.68	116,507	10,290	–	–	–
ceramics	0.98	9,259	818	–	–	–
bricks	4.09	26,607	2,350	3.11	3,432	303
clay	37.63	18,816	1,662	11.50	–	–
hemp, jute, straw, wattle	15.42	6,487	356	8.27	1,985	109
cellulose, wood fibres	3.48	66,471	5,499	–	–	–
timber	49.24	520,998	29,278	28.22	74,504	5,609
other	1.63	18,243	1,666	0.91	–	–
total	*391.14*	*1,308,184*	*104,571*	*104.00*	*82,715*	*6,267*
total per m²GIA	*1.03*	*3,452*	*276*	*0.99*	*788*	*59.69*

In the case of Villa Strohbunt, values inputted actually refer to industrial origin materials; for instance, felling with an electric chainsaw and kiln-drying have been included in the calculation of the GWP of timber. This certainly affects the results negatively, i.e., they are overestimated.

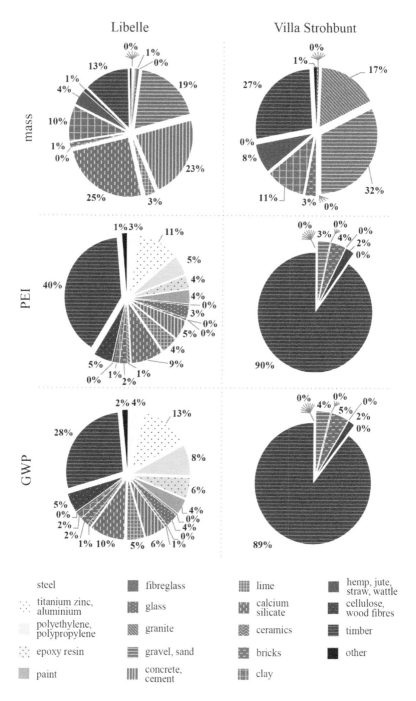

Figure 3.1 Libelle mass, PEI and GWP and Villa Strohbunt mass, PEI and GWP

Table 3.1 and Figure 3.1 contain the results for both buildings. The total CO_{2eq} emissions entailed by the construction of Libelle are 104.57 t (0.28 tCO_{2eq}/m^2GIA and 10.46 tCO_{2eq}/person); of Villa Strohbunt are 6.27 t or 0.06 tCO_{2eq}/m^2GIA and 1.04 tCO_{2eq}/person.

So far, positive GWP values have been used for vegetal-origin elements. If we now subtract the CO_2 sequestered by the living plants, Libelle's GWP becomes 47.05 tCO_{2eq} (i.e. 0.12 tCO_{2eq}/m^2GIA and 4.71 tCO_{2eq}/person) and Villa Strohbunt's −23.02 tCO_{2eq} (−0.22 tCO_{2eq}/m^2GIA and -3.84 tCO_{2eq}/person).

It might also be noteworthy to observe that the ratio GWP/building weight [$kgCO_{2eq}$/kg] is 0.27 for Libelle, and 0.06 (−0.22) for Villa Strohbunt.

3.4 PEI and GWP of technical services

With technical services we mean all the heating, cooling, ventilation, water and electrical systems. These are not included in most LCAs of buildings due to lack of complete information; moreover, products are complex in terms of materials, which makes analyses very difficult. Many analyses on individual elements are indeed available, especially for solar panels, for instance Ardente et al. (2005), and heat pump, for instance Koroneos (2016), but rarely on the complete circuit. No studies have been found as regards electrical systems.

In the case of Libelle it was possible to carry out the same analysis performed on structural materials as regards the technical services thanks to a bill of quantities complete with all the elements used in the building. Unfortunately, it was not possible to trace comparable quality data for Villa Strohbunt due to its particular design and construction circumstances. For this reason it would be risky to hypothesise the design of the technical installations, which are reduced to the minimum necessary and therefore deviate from average buildings.

For domestic hot water (DHW) and space heating Libelle is equipped with three systems which are integrated through a 12.4 m^3 storage tank: a solar circuit with a 100-litres expansion tank; a geothermal circuit; and a hydro stove with two 800-litres expansion tanks.

The solar circuit consists of a solar surface of 3 × 21 m with an inclination of 75° to the south, with mineral wool insulation, highly selective-coating copper absorber (95% absorption), and double anti-reflective tempered glass sealed with EPDM.

The geothermal circuit consists of a heat exchanger made of PEX pipes laid spiral-wise from the inside to the outside in the filling layer of the ground floor.

The stove circuit consists of a 14.9 kW inverted flame firewood hydro stove and radiators and towel-warmers in the bathrooms.

From mid-February to mid-November, the solar system is enough to supply hot water and space heating. At winter's peak, it is replaced by the stove. Heat distribution and solar system control are automatic.

The drinking water system comprises polypropylene and HD-polyethylene piping arranged in three circuits: one to feed the storage tank; one for cold water; and one for hot water.

The sewage system flows greywater into the ecovillage treatment plant (reed bed). There is no production of blackwater as toilets are dry separation toilets, each of which is equipped with an air extraction and filter system.

There is a central ventilation system with a heat recovery unit. Each living room and bedroom has a vent to supply fresh air. Kitchens, bathrooms and part of the corridors are provided with filter and extraction air vents. Libelle is divided into four ventilation zones: west ground floor, east ground floor, west first floor, east first floor. In each zone the air flows from the living spaces to a corridor, a bathroom or a kitchen through extraction vents or passages in walls and doors and is then expelled. External air is filtered through two coarse filters G3 and G4 (gross filters, average arrestance Am (%): $80 \leq$ Am < 90) and preheated by the heat recovery unit.

The propane gas system consists of a circuit supplied by two gas cylinders (33 kg capacity) and serving the three kitchens in the building.

The electrical system is provided with a single meter for the entire house, which is located on the ground floor. There is a distributor for each of the four zones of the house, with a fault current switch (isolation) and an automatic switch (overload) for each room.

Each living room has a TAE telephone socket and a TV connection (connected to the satellite dish); each room is provided with a UAE Ethernet socket.

The calculations were performed using the same methods used for the structural building elements in Chapter 3.3. It was difficult to identify the constituent materials of many elements, especially of the electrical system. This, in fact, consists of small highly composite elements. Technical data sheets of electrical products rarely declare the materials that make them up, and even less often their percentages. Furthermore, studies about these elements are rare. Therefore, when we could not find precise information, the quantities of the main materials have been obtained from their dimensions: for example, for electric cables the cable section was used as a basis. However, it was possible to assimilate some products used in Libelle, such as switches, to ones present in PEP-ECOPASSPORT[1] (PEP-ECOPASSPORT) with an LCA from which the percentages of constituting materials were obtained. In no single case was exactly the same product found, but similar products were, and almost always from the same manufacturer.

However, it was not possible to identify the real composition of large complex elements, such as the stove, the kitchen stove and the mechanical ventilation heat recovery (MVHR) unit.

For the first two, reference was made entirely to the main material (stainless steel), while for MVHR to the two materials mentioned in the technical sheet: galvanised steel plate for the housing and plastics for the heat exchanger.

Finally, as regards to the solar panels, reference was made to values found for one flat-plate collector of 1.9 m² (Kalogirou 2004), from which we proportionally calculated EE and EC for the amount installed on Libelle's roof, that is 63 m². Table 3.2 and Figure 3.2 contain the results.

It is interesting to note that technical services are just 2% of the total mass of the building, but their impact in terms of primary energy and GHGs are 19% and 24%, respectively. This is due to the fact that technical systems elements are mainly composed by metals and plastics with very high PEI and GWP values. If negative values are admitted for GWP, the weight of technical services increases to 40%: this is due to the almost complete absence of vegetal-origin materials in technical services parts.

Table 3.2 Mass, PEI and GWP of Libelle technical services

Material	Weight [t]	PEI [MJ]	GWP [kgCO$_{2eq}$]	GWP [kgCO$_{2eq}$] (negative values admitted)
iron, steel, cast iron	3.84	105,131	10,434	10,434
silver, zinc, aluminium	0.03	4,980	494	494
copper	0.96	40,416	4,011	4,011
brass	0.11	4,794	476	476
polyethylene, polypropylene and other plastics	1.30	118,805	14,855	14,855
epoxy resin	0.00	20	2	2
paint	0.04	2,778	248	248
fibreglass	0.00	33	3	3
glass	0.52	8,032	709	709
ceramics	0.06	1,879	166	166
hemp	1.70	8,500	467	467
timber	0.48	4,752	261	− 591
other	0.30	12,807	870	− 143
total	*9.36*	*312,928*	*32,996*	*31,531*
total per m²GIA	*0.02*	*826*	*87*	*83*

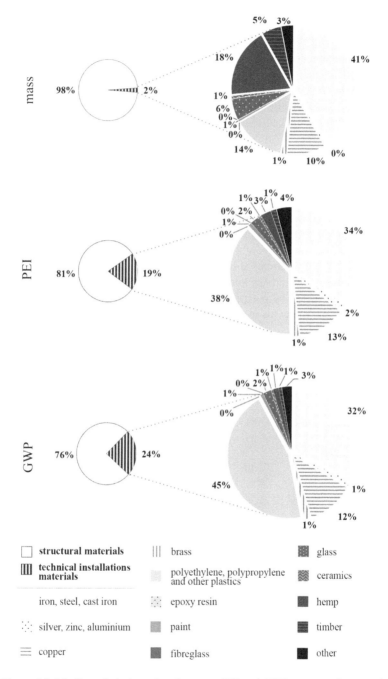

Figure 3.2 Libelle technical services by mass, PEI and GWP, compared to grand totals

3.5 Comparisons

All comparisons that follow do not take into account the technical services.

3.5.1 Comparison with Wegmann-Gasser house

An interesting comparison can be made with a similar building: Wegmann-Gasser house in Glarus, Switzerland, designed by architect Werner Schmidt with a similar technology (load-bearing straw bale construction, with timber structures for floors and roof) (Bocco Guarneri 2013): see Table 3.3 and Figure 3.3.

If we consider the impacts per square metre, it is immediately apparent that Villa Strohbunt, thanks to the constructional choices (reuse of some elements and almost exclusive use of natural materials processed without the use of machinery) has the lowest PEI and GWP. Wegmann-Gasser House has the highest figures per square metre, because though the absolute amount of resources and energy is much lower (40% less in weight and 30% in PEI than Libelle), its floor area is about 50% less than that of Libelle.

The differences remain almost unchanged if we consider the impacts per inhabitant. Villa Strohbunt remains the building with the best results, also thanks to its high occupation rate: 0.057 persons per m^2 (Libelle is 0.026 persons per m^2 and Wegmann-Gasser 0.028 persons per m^2).

It is interesting to note that in case negative values are admitted for GWP, the gap between Wegmann-Gasser house and Libelle increases: the percentage of vegetal materials in the Wegmann-Gasser house is about the double of that of Libelle, so in the first building the GHGs emissions due to other materials are compensated and exceeded.

3.5.2 Comparison with average buildings

Another interesting comparison can be drawn between Sieben Linden buildings and average residential buildings. Unfortunately, no study has been found that provides the average environmental impact of buildings at the national level (Germany) or at the European level. Most studies, in fact, provide collections of case studies, selected according to different criteria. We used two works as references, which allow some extent of comparability: Dixit (2017) shows PEI values for residential buildings by continent and calculates the average value for different construction systems – timber, reinforced concrete and steel. Birgisdóttir (2017) discusses eighty-one case studies worldwide, among which only eighteen were selected because the others were office buildings or schools, had different boundaries, or data were incomplete. We used the mean PEI value calculated from the relevant

Table 3.3 Comparison of Mass, PEI and GWP values of Libelle, Villa Strohbunt and Wegmann-Gasser house

	Weight	PEI			GWP			GWP (negative values admitted)		
	$[t]$	$[GJ]$	$[GJ/m^2]$	$[GJ/p]$	$[tCO_{2eq}]$	$[tCo_{2eq}/m^2]$	$[tCO_{2eq}/p]$	$[tCO_{2eq}]$	$[tCo_{2eq}/m^2]$	$[tCO_{2eq}/p]$
Villa Strohbunt	104.00	82.72	0.79	13.79	6.27	0.06	1.04	−23.02	−0.22	−3.84
Libelle	391.14	1,308.18	3.45	130.82	104.57	0.28	10.46	47.05	0.12	4.71
Wegmann-Gasser	239.32	964.79	5.33	192.96	65.59	0.36	13.12	−27.03	−0.15	−5.41

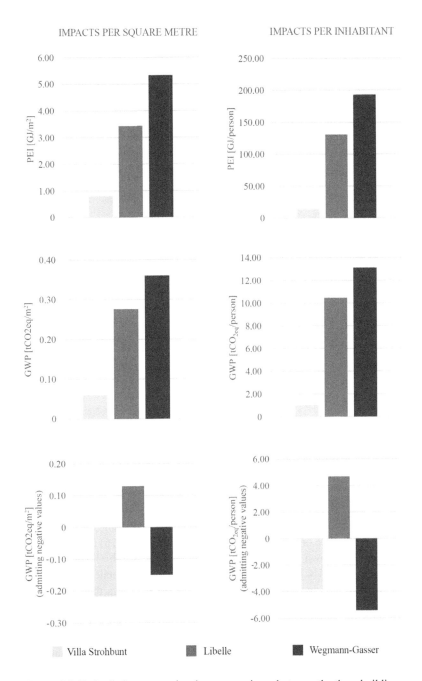

IMPACTS PER SQUARE METRE

IMPACTS PER INHABITANT

Villa Strohbunt   Libelle   Wegmann-Gasser

Figure 3.3 Embodied energy and carbon comparisons between the three buildings

Table 3.4 Comparison between Libelle, Villa Strohbunt and the mean values for
the most common construction systems

Construction system	PEI mean values [GJ/m²]	GWP mean values [tCO₂ₑq/m²]
timber	2.20	0.20
reinforced concrete	5.77	0.22
steel	10.64	0.28

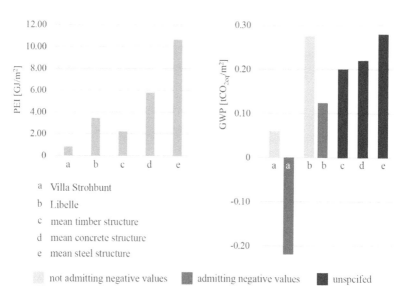

Figure 3.4 Comparison between Libelle, Villa Strohbunt and the mean values for
the most common construction systems

cases extracted from both studies as a benchmark (see Table 3.4), while the
mean GWP value only data contained in Birgisdóttir (2016) could be used.

Again, as shown in Figure 3.4, Villa Strohbunt has the lowest embodied energy, at 64% less than mean timber buildings; Libelle's PEI is 57%
higher than average. Villa Strohbunt's GWP is 70% less than mean timber structures; Libelle's is significantly (27%) above the mean value. The
results obtained present inconsistencies, namely as Libelle unrealistically
compares badly with the average. Yet, if we admit negative values, Villa
Strohbunt's GWP becomes –110% less than mean timber buildings and that
of Libelle 38% lower. As the sources do not specify whether they did admit
negative values for GWP, we have to confine ourselves to observing that

our references cannot really be used as benchmarks. Some reasons for this and other possible inconsistencies will be discussed in the next paragraph.

3.6 Discussion of Embodied Energy and Embodied Carbon results

In Chapter 3.5.1, Villa Strohbunt's values are far below the other two buildings for all indicators. This is due to the fact that few materials, in weight and variety, have been used in its construction, most of which required simple processing without the use of machinery. Another relevant factor is that many elements were salvaged and reused.

In terms of values per square metre, it is clear that in spite of similar building technology and energy performance, Libelle has a smaller impact on the environment than Wegmann-Gasser house. It should be remarked that values per inhabitant ought to be considered as well, because it would be irresponsible to build huge high-performance houses for a small number of residents and heat their whole space. Indeed, the number of residents affects the results, although it changes (or may change) over time.

Unfortunately, all comparisons had to be made without taking into account the technical services, due to lack of data. From the analysis conducted on Libelle, however, it can be remarked that their contribution to PEI and GWP values is all but negligible, at about 20% of the total. Life-cycle inventories should fully cover not only buildings' structures but also their technical services. (At the moment, their impact is usually assessed in the use phase only). For this purpose, bills of quantities should include services, which is not the case, as the working project of these is the responsibility of several professionals, different from the designer of the building. Even more relevant is that much more information would be needed in product data sheets and more specific studies on individual products would be necessary, especially the electrical ones: however reasonable the assumptions one can make, their complexity makes it very difficult to describe them correctly if accurate information from manufacturers is lacking.

On the other hand, the results obtained from the comparison with European average values (Chapter 3.5.2) cannot be considered rigorous; rather they show how problematic it is to return a result that is truly comparable with those obtained in other studies.

EE values reported by Dixit (2017) vary up to 50%; he explained such range with methodological issues, such as differences of system boundary and calculation method, and also with data quality issues. These inadequacies affect the results; for instance they are very seldom so transparent to allow disaggregation and therefore made consistent in terms of system boundaries (actually, an accurate description of the system boundaries was sometimes lacking altogether). Some studies take into account transport operations, others the EE of building products only. As mentioned above (Chapter 3.2), in our own calculations we had to stay within phases A1–A3 of an LCIA (that is, from raw materials supply to building materials production) as we did not precisely know the distance travelled to the building site (phase A4) nor could we in detail describe the assembly operations at the building site itself (phase A5). Therefore, we picked from Birgisdóttir (2016) only the cases with the same system boundaries as ours; although the EE values provided by Dixit (2017) were reduced by 16% to normalise them before calculating the mean values. In fact, the author states that on average transport and construction contribute by 6% and 10% respectively to the total EE values.

Another critical issue is the creation of really exhaustive inventories. In the present study, for example, it was not possible to consider the changes intervened during construction. Some elements not fully described in the bill of quantities were (with some approximation) equated to products on the building market (waterproofing sheaths), or hypothesised (reinforcement of internal doors lintels). The studies selected for comparison do not provide information so detailed as to allow a safe evaluation of the approximations made.

These can therefore be large, as in the case of EE calculation. The main calculation methods are input output (IO)-based, process-based, hybrid. The process method takes into account the energy and material inputs one by one, but sometimes remains incomplete due to the difficulty in obtaining the data. The IO method is based on site book-keeping: payments to buy materials are translated into energy. The hybrid method combines the previous two. According to Optis (2010), 78% of the studies do not provide information on the method used.

Another problem issue is the unit: for example, sometimes only EE per square metre and per year is provided, without declaring the duration of the assumed lifespan of the building nor the absolute EE value (moreover, it is rarely specified if GIA, GFA, or other surface areas were considered). In the case studies collected by Birgisdóttir (2016) it has not always been possible to identify the reference surface, while in Dixit (2017) it is never specified. This means that values we used for comparison might be underestimated.

According to Optis (2010), about 20% of the studies do not adequately reference data sources. The sources used can be primary (software datasets) or secondary (other studies). Most databases may not be complete or accurate; in fact they use different criteria of data selection, they are seldom annually updated, and values for some unconventional materials are lacking, in particular vegetal-origin materials. Sometimes, when some such materials are covered, databases seem to be biased against them. The lack of values can affect the consistency of the assessment of buildings mainly made of plant materials (see Chapter 5.1).

As regards GWP, average values are quite low compared to Libelle, arguably for the same reasons stated previously. Furthermore, they might have been influenced by the different national energy mixes (the same building would present different EC values according to the country where it is built) and the possible acceptance of negative values in the calculation. This very relevant issue is not specified in the reference studies, but affects the comparison (see Figure 3.4) so much as to make it almost meaningless.

In conclusion, comparisons between results from different studies can be considered reliable insofar they share assumptions and procedures. Until the problems discussed in this section are at least partially solved, it is not even possible to determine a benchmark.

3.7 Ecological footprint of Sieben Linden buildings

We assess here the EF of the resources employed in the same two buildings studied in Chapters 3.1, 3.2 and 3.3, taking into account the most significant impacts on bioproductive areas. This analysis was carried out according to the EF component method; the boundaries are from cradle to gate, as for PEI and GWP, but do not take into account the technical services. Conversion factors (yield, equivalence factor) are the same used in Chapter 2.

Table 3.5 shows that absolute EF values of Libelle are higher than those of Villa Strohbunt, which is not very telling as Libelle is a larger building. However, if we consider the unitary EF [gha/m²GIA], the differences between the two buildings become much smaller: in particular, Libelle's energy land is 3.5 times larger than Villa Strohbunt's, whereas the forest land of the latter is 1.4 times larger than Libelle's, in fact the amount of vegetal-origin building materials in Villa Strohbunt is 0.25 t/m²GIA while in Libelle is 0.14 t/m²GIA.

Unfortunately, no other studies have been found on the EF of strawbale buildings or ecovillage buildings for comparisons. Therefore, the

Table 3.5 Libelle and Villa Strohbunt EF

	Libelle			Villa Strohbunt		
	gha	*gha/m²*	*%*	*gha*	*gha/m²*	*%*
energy land	26.84	0.07	23.38	1.61	0.02	4.49
forest land	87.21	0.23	76.57	34.18	0.33	95.45
built-up land	0.06	0.00	0.05	0.02	0.00	0.06
total	*114.12*	*0.30*	*100.00*	*35.81*	*0.34*	*100.00*

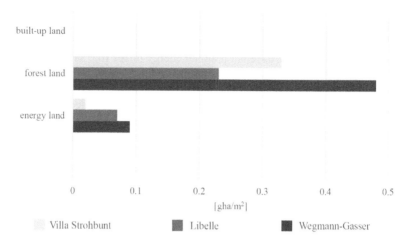

Figure 3.5 Comparison of Libelle, Villa Strohbunt and Wegmann-Gasser EF

Wegmann-Gasser house was maintained as the benchmark, as available data allowed us to calculate its EF at 0.57 gha/m²GIA (0.09 gha/m²GIA energy land + 0.48 gha/m²GIA forest area: see Figure 3.5).

Wegmann-Gasser house's values are the largest for each component of the EF; the total value exceeds that of Libelle by 45.6%, and of Villa Strohbunt by 36.8%, confirming the results of the PEI and GWP analyses. But as opposed to these, it is now Villa Strohbunt that is the building with the second-largest impact.

The three components of the EF of buildings have a very different impact on the final result. The built-up land can be considered negligible, while the forest area scores highest. This is because the buildings analysed employ large amounts of vegetal materials. Paradoxically, this variable would take lower or even zero values in buildings using other construction systems such as reinforced concrete. In fact, the EF measures the consumption of renewable resources, i.e. the land in terms of resources which the planet can regenerate. All non-renewable resources, i.e. those resources that require geological time for their reproduction, are excluded from the calculation. It is only considered the impact produced by the processing and extraction of these resources that is accounted for – it is expressed in CO_2 emissions and is converted into land for energy, i.e. the forest land necessary for their absorption.

This indicates that this method is inadequate to weigh non-biological components and it merely converts the PEI attributed to them in equivalent hectares of "energy land." However, the ecological footprint method can provide a more complete result than the PEI assessment, because it does not only include the energy used in the production processes but also the consumption of materials (when of vegetal origin).

Note

1 PEP ecopassport is an non-profit association with the aim to develop internationally the environmental declaration program PEP ecopassport concerning electrical, electronic and HVAC (heating, ventilation, air-conditioning, refrigeration) products.

References

Ardente, Fulvio et al., "Life cycle assessment of a solar thermal collector: sensitivity analysis, energy and environmental balances", *Renewable Energy*, 30, 2, 2005, pp. 109–130. doi:10.1016/j.renene.2004.05.006

Berge, Bjørn, *The Ecology of Building Materials*, Oxford: Architectural Press, 2009.

Birgisdóttir, Harpa et al., "IEA EBC annex 57 'evaluation of embodied energy and CO$_{2eq}$ for building construction'. Case studies demonstrating Embodied Energy and Embodied Greenhouse gas Emissions in building", *Energy and Buildings*, 154, November 11, 2017, pp. 72–80. doi:10.1016/j.enbuild.2017.08.030

Bocco Guarneri, Andrea, *Werner Schmidt Architect: Ecology Craft Invention*, Vienna: Ambra Verlag, 2013.

Destatis, *Energy Use of Local Units in Manufacturing*, 2015 [online]. Available from: www.destatis.de/EN/FactsFigures/EconomicSectors/Energy/Use/Tables/EconomicBranch.html [last viewed Sept. 2018].

Dixit, Manish K., "Life cycle embodied energy analysis of residential buildings: A review of literature to investigate embodied energy parameters", *Renewable and Sustainable Energy Reviews*, 79, 2017, pp. 390–413. doi:10.1016/j.rser.2017.05.051

Fachagentur Nachwachsende Rohstoffe e. V. (FNR), *Studie zur Markt- und Konkurrenzsituation bei Naturfasern und Naturfaser-Werkstoffe*, Gülzower Fachgespräche 26, Gülzow: FNR, 2008.

Fachagentur Nachwachsende Rohstoffe e. V. (FNR), *Marktübersicht Dämmstoffe aus nachwachsenden Rohstoffen*, Gülzow-Prüzen: FNR, 2014.

Hammond, Geoffrey; Craig Ian Jones, "Embodied energy and carbon in construction materials", *Proceedings of the Institution of Civil Engineers: Energy*, 161, 2, 2008, pp. 87–98. doi:10.1680/ener.2008.161.2.87

Hammond, Geoffrey; Craig Ian Jones, *The Inventory of Carbon and Energy (ICE)*, Bath: University of Bath and the Building Services Research and Information Association, 2011.

Hegger, Manfred et al., *Energy Manual: Sustainable Architecture*, München: Edition Detail, 2008.

Juhrich, Kristina, CO_2 *Emission Factors for Fossil Fuels*, Dessau-Roßlau: Umweltbundesamt, June 2016.

Kalogirou, Soteris A., "Environmental benefits of domestic solar energy systems", *Energy Conversion and Management*, 45, 2004, pp. 3075–3092. doi:10.1016/j.enconman.2003.12.019

Kono, Jun; York Ostermeyer; Holger Wallbaum, "The trends of hourly carbon emission factors in Germany and investigation on relevant consumption patterns for its application", *The International Journal of Life Cycle Assessment*, 22, 10, 2017, pp. 1493–1501. doi:10.1007/s11367-017-1277-z

Koroneos, Christopher J.; Evanthia A. Nanaki, "Environmental impact assessment of a ground source heat pump system in Greece", *Geothermics*, 65, 2017, pp. 1–9. doi:10.1016/j.geothermics.2016.08.005

Optis, Michael; Peter Wild, "Inadequate documentation in published life cycle energy reports on buildings", *The International Journal of Life Cycle Assessment*, 15, 7, 2010, pp. 644–651. doi:10.1007/s11367-010-0203-4

PEP-ECOPASSPORT, *Find a PEP* [online]. Available from: www.pep-ecopassport.org/find-a-pep/ [last viewed Sept. 2018].

Stanellé, Chironya; Iris Kunze (eds.), *20 Jahre Ökodorf Sieben Linden*, Poppau: Freundeskreis Ökodorf, 2017.

TU Berlin, *Stroh im Haus, statt Stroh im Kopf*, Berlin: Energieseminar – Institut für Energietechnik (TU Berlin), 2006.

van Dam, Jan E.G.; Harriëtte L. Bos, *The Environmental Impact of Fibre Crops in Industrial Applications*, 2004 [online]. Available from: www.researchgate.net/publication/40110933_The_environmental_impact_of_fibre_crops_in_industrial_applications [last viewed Sept. 2018].

Wiegand, Elke; Martin Stengel; Dirk Scharmer, *Ecovillage Sieben Linden with Straw bale construction*, Paper compiled as an entry to the World Habitat Awards 2006 Building and Social Housing Foundation competition, Sieben Linden, 2006.

4 Comparing daily impact and construction impact

Martina Gerace and Susanna Pollini

An overall assessment of the environmental impact of the ecovillage is proposed here, which takes into account both the consumption of Sieben Linden residents and the construction of buildings.

4.1 Recurring footprint and building footprint

The EF associated with the daily activities of residents has been calculated for the entire ecovillage for 2014 in Chapter 2. It is 389.18 gha, and 3.06 gha/person (public services included). [1]

The estimate of the EF of buildings was based on the figures of Libelle, as it is satisfactorily representative of average Sieben Linden residential buildings in terms of construction techniques and surface/residents ratio (see Chapter 3.7).

An annual construction index (1.06, based on the average amount of square metres built yearly) was then introduced to calculate the yearly average EF due to construction activities (Ardente, Fulvio et al.,). The average annual EF of buildings is thus 81.19 gha and 0.64 gha/person. If we add this to [1] here, the impact of construction accounts for 17% of the total. In spite of the low impact of the ecovillage's buildings, their share is relatively high – in fact, also the EF of everyday lifestyle is low.

The EF associated with the construction of buildings is not generally considered in the calculation of the footprint of sub-national areas. This is one of the aspects that makes it difficult to compare Sieben Linden EF with national EF. In fact, the standard method for calculating the national EF considers a wide variety of impacts, including those associated with the construction of all private and public facilities. To compare the local EF with the national EF it would be necessary to align with this method, and calculate every single entry in the EF of sub-national areas assessment.

On the other hand, we cannot accept the amortisation of the environmental impacts of the construction activity. This would be against one of the fundamental assumptions of the EF methodology (see Chapter 2.1.2), and anyway the actual service life of a building is unpredictable and much more dependent on social (real estate market, fashion, war . . .) and physical phenomena (fires, earthquakes . . .) than on technical ones; indeed, conventional service life spans (assumed as significantly smaller than 100 years) seem to promote focussing on recurrent energy performance rather than on durability and low EE.

This said, the figure 0.64 gha/person we calculated for Sieben Linden does not seem completely unlikely if compared with an average national value of 0.11 gha/person for residential buildings (as from EUREAPA), as Sieben Linden is growing at a much faster rate than Germany – although with low-impact materials and technologies. On the other hand both at the local level and at the national level we used the same figure as of EUREAPA for infrastructure construction (0.31 gha/person).

Table 4.1 and Figure 4.1 show a comparison between the overall EF of the ecovillage (including individual consumption, national services, and construction) and the average German EF.

Table 4.1 Comparison between Sieben Linden footprint and German average footprint

	EF [gha/person]			
	individual consumption	*construction**	*public services***	*total*
Sieben Linden (2014)	1.74	0.95	1.01	*3.72*
German average (2013)	n/a	n/a	n/a	*5.47*
German average as of 2004 (EUREAPA)	4.64	0.42	1.01	*6.04*

*includes residential and infrastructure construction
**includes public administration and leisure

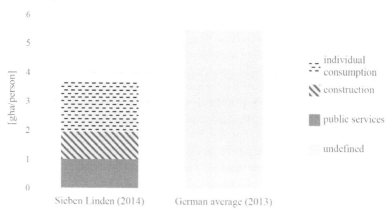

Figure 4.1 Comparison between Sieben Linden footprint and German average footprint

Even including the impact of construction activities, which leads to a 21% increase in value [1], the EF of Sieben Linden is 32% lower than the German national average (see Table 4.1 and Figure 4.1).

4.2 Energy consumption and GHG emissions: operating vs. embodied impact

As meters and firewood management are centralised in the ecovillage, it was not possible to stipulate how much energy is consumed by each building. Therefore, for Villa Strohbunt the average per capita value, 16.54 GJ/person*year, was multiplied by the number of residents (this energy amount includes gas, electricity and firewood consumption). Such average value was calculated considering both residential and communal spaces as it was not possible to tell the two functions apart. However, the amount of firewood burnt in the Libelle was known (0.6 raummeter per person), so in this case an operational energy value of 9.51 GJ/person*year was obtained.

Libelle's PEI is considerably higher than operational energy: it takes almost fourteen years of building use to equal it. As we have said previously in Chapter 4.1 we assume that Libelle can be considered an acceptable proxy for the average Sieben Linden building. However, the energy values shown in Table 4.2 and Figure 4.2 cannot be directly transferred to the whole ecovillage, as it should be borne in mind that many residents live in trailers rather than buildings.

In Figure 4.2, we have therefore proposed a weighted version of this same analysis: the total operational energy of Sieben Linden was obtained by multiplying the figure per person by the number of people living in residential buildings; whereas the total PEI of Sieben Linden buildings is the sum total of 1) Libelle's unit PEI multiplied by the built-up residential area; and 2) Libelle's unit PEI multiplied by the amount of communal built-up spaces weighted on the number of people living in buildings (that is, Sieben Linden population minus the people living in trailers). Refurbished buildings

Table 4.2 Energy and emissions of buildings

	Energy		Emissions	
	operational energy [GJ/year]	PEI [GJ]	operational emissions [tCO$_{2eq}$/year]	GWP [tCO$_{2eq}$]
Libelle	95.1	1,308.2	1.74	104.57
Villa Strohbunt	99.2	82.7	1.72	6.27

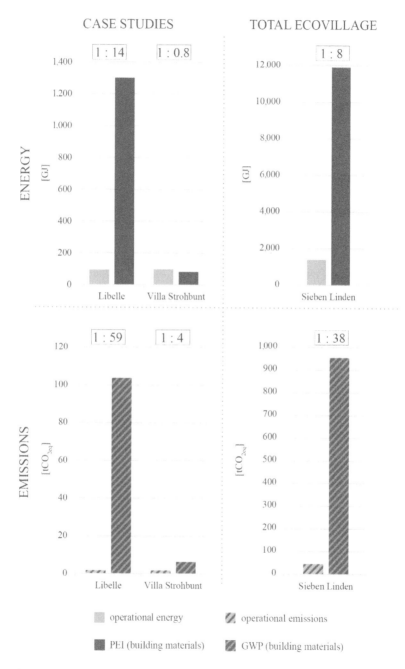

Figure 4.2 Comparison of operational energy and PEI of buildings and of operational emissions and GWP of buildings

were assigned PEI = 0 as they are reused structures, and the impact of their upgrading has been assumed as very low.

The estimated total PEI is now a bit more than eight times larger than the overall operating consumption – that is, it would take more than eight years to use as much energy as in the making of the buildings. This value is lower than that obtained for Libelle, as Libelle's firewood consumption is lower than Sieben Linden average.

A similar procedure was repeated for GHG emissions (see Table 4.2 and Figure 4.2). In this case too, operational emissions values were available at the whole ecovillage level; therefore we assigned to Villa Strohbunt the average per capita value (0.29 tCO_2/person*year) multiplied by the number of residents, while for Libelle we obtained 0.17 tCO_2/person*year on the basis of the amount of firewood actually burnt.

In this case, the difference between operational and building phases is much larger. For Libelle, it takes fifty-nine years for the two values to equalise. This depends on the energy sources used: the production of industrial building materials employed the German energy mix, while the ecovillage runs on nearly zero-emission renewable energy.

Comparing the values for the whole ecovillage (Figure 4.2), it would take thirty-eight years for the GHG emitted to operate the buildings to equal the amount released when the building materials were produced. Again, this value is lower than that calculated for Libelle, because of its lower-than-average firewood consumption.

5 Final remarks, recommendations and perspectives

Andrea Bocco

5.1 On the methodologies used

Ecological Footprint appears as the most consistent method for lifestyle impact analysis, especially considering its ability to include many different aspects of everyday life (e.g. mobility, food, operational energy in buildings, etc.). The main commonly recognised strengths of EF are: "(a) its ability to condense the size of human pressure on different types of bioproductivity into one single value, (b) the possibility to provide some sense of over-consumption, and (c) the ability to communicate results to a wide audience" (Wiedmann 2010).

On the other hand, such method implies several limitations:

> First of all, EF only focuses on renewable resources consumption and does not capture the full range of environmental impacts, such as those arising from acidification, eutrophication, ecotoxicity, human toxicity, etc. Those impacts imply processes that may irreversibly damage bioproductive capacity, e.g. by reducing ecosystem services, affecting nutrient cycles, impacting biodiversity.
>
> (Castellani 2012)

Less impactful practices like organic agriculture are conformed to conventional practices: this can imply an overestimation of food EF when such practices are adopted, as in the case of Sieben Linden.

Moreover, in EF calculations yields are obtained from current farming, fishing, forestry etc. practices regardless of their sustainability, that is to say, their aptitude to leave the biocapacity unchanged (Wiedmann 2010; Castellani 2012).

In addition, the EF methodology does not account for the possibility of multiple functions of an ecosystem (e.g. it assumes that a given land surface cannot simultaneously produce timber and absorb carbon) and this may result in a larger area of land being shown as required than actually necessary (Castellani 2012). Nevertheless, it is important to remark that, in contrast with other methodologies such as Carbon Footprint, EF does not account for carbon emissions derived from biotic sources combustion. This means that, in the case of firewood, carbon sequestered during the lifetime of trees compensates the carbon dioxide emitted during combustion. Firewood used for heating in Sieben Linden generates considerable effects in terms of forest area needed to grow it, but none in terms of forest area needed to absorb emissions.

Given the complexity of the EF calculation, some simplifications are necessary. One of the most relevant is the use of average world yield factors instead of local yields. This allows to easily compare results of different studies worldwide; but on the other hand entails difficulties in comparing with the local biocapacity as it just allows comparison with the world average biocapacity (see Chapter 5.2). As Sieben Linden is committed to achieving self-sufficiency within the limit of the sustainable use of local natural resources (such as firewood), comparison with local yields would be relevant.

Despite the consistent efforts made in order to standardise the EF methodology, a common method for sub-national EF accounting has not yet been developed (see Chapter 2.1.2) (Lewan 2001). The most difficult aspect of such standardisation concerns the source and quality of data on consumption. In fact, while at the national level a top-down approach (which means, based on national trade statistics and energy budgets) is used, the bottom-up approach (widely applied in sub-national analyses) inevitably leads to discrepancies between single studies (see Chapter 2.5.4). Moreover, the provision of data at the local level is a considerably time-consuming operation, and this can reduce the chances of monitoring long-term progresses.

As explained in Chapter 3.2, ICE database was chosen to calculate the EE and EC of buildings because it is open access and readily understandable even for inexperienced users, in spite of being based on verifiable sources, and, last but not least, for being one of the most reputed tools of its kind existing in the English language. However, when picking a database, one inevitably needs to embrace assumptions at its very roots, some of which users are not immediately aware of, and even fewer can modify to adapt to their own special situations. In our case this would have been particularly relevant as (similarly to what differentiates organic from conventional agriculture) the conventional process from felling a tree to having a dimensional lumber element ready for use at the building site implies a much greater number of operations (including sawmilling, packaging, etc., and several transportations) than what actually happens in an "alternative" build where some of these operations are skipped altogether and others are of very limited extent and sometimes performed without powered tools. In other words, this means that referring to a database that cannot – inevitably – cover the full range of fine-grain variations in how materials are actually processed and just provides values for conventional manufacturing will result in an overestimation of the environmental impacts, which is certainly the case here for Villa Strohbunt and, to a much lesser extent, for Libelle and Wegmann-Gasser houses as well. Mainstream databases appear ill-suited to correctly describe the impacts associated with labour-intensive building operations making use of locally sourced, low-technology, organic-grown building materials which are quite common in "green" buildings, and particularly self-built ones.

We still need to add that even within such limitations, a sometimes-disorienting discrepancy between values provided by respected databases exists. For instance, had we used Ökobaudat database (Oekobaudat 2017) for our calculations the results would have probably been better (i.e., showing lower impact) than with ICE, at least as long as GWP is implied, while PEI would have possibly resulted higher. These divergences are not just due to obvious differences in the national (UK and German) "energy mixes" and industrial manufacturing processes, but also because of dissimilar methodology and assumptions in the assessment procedures, particularly (but not only) for what concerns the calculation of how much CO_2 is locked up in a vegetal-origin building material. (Woolley 2013). (A discussion of the discrepancies of different approaches to LCA applied to buildings can be found in Chau et al. (2015). Pomponi (2016) shows how different can be the measures suggested to mitigate the negative impacts, according to the methodologies and assumptions taken by different authors.) The discrepancy between the data contained in different databases can be so large as to sometimes induce questioning their sheer reliability. In the case of Libelle, we obtained 3.45 GJ/m^2GIA and 0.12 tCO_{2eq}/m^2GIA making use of the ICE database, but we would have found 5.39 GJ/m^2GIA (+56%) and -0.01 tCO_{2eq}/m^2GIA (–108%) had we referred to Ökobaudat. A large part of such non-negligible divergence is due to the different environmental profiles of timber and timber-based products in the two databases. The discrepancy between databases has not yet been adequately investigated, probably because each of them is so far used almost exclusively within the country where it was originated, and because those who perform LCIA assessments usually rely on proprietary software and seldom calculate the environmental impact of a building summing up the single components manually. A European database covering a full range of building materials and equipped with conversion factors for different countries would greatly help.

One more caveat concerns the functional unit considered: usually impacts of buildings are referred to by the square metre (e.g.: MJ/m^2, $kgCO_{2eq}/m^2$)

but one should attentively check that the surface area is consistently defined: in buildings where the envelope is insulated with natural materials such as straw bales, the thickness of perimeter walls can be significant, affecting the gross floor area. There can be a notable difference between values referred to by the m^2GFA and the m^2GIA: obviously, the numerator is the same while the denominator is higher in case GFA is considered, and therefore values referred to by the m^2GFA are smaller. For instance, the 3.45 GJ/m^2GIA and 0.12 tCO_{2eq}/m^2GIA we obtained for Libelle translate into 3.07 GJ/m^2GFA and 0.11 tCO_{2eq}/m^2GFA respectively. For this reason, results shown in Chapter 3.3–3.5 need to be compared with results referred to the m^2GIA only, lest be unfair to the cases here analysed and, in general, to buildings with thick perimeter walls.

Finally, we would like to underline how crucial is the first step of any environmental assessment, inventorying. For everyday activities covered in Chapter 2.4 we could rely on quite robust data, most of which were routinely collected by the Sieben Linden community even before this study started. On the other hand, it is not their common practice to record energy and material fluxes on building sites (and this is why we recommend to improve this: see Chapter 5.3), and therefore we had to check sometimes incomplete or (in the case of Villa Strohbunt) altogether missing bills of quantities. Another method to get to a thorough inventory of the materials that make up a building would have consisted in creating an accurate BIM model that was fully integrated, or at least fully interoperable, with a reliable LCIA tool (König et al. 2010:78 ff.; see also, for instance, Capper et al. 2012). (A BIM model would not contain the building materials that are wasted in the construction process, though.) However, while the environmental impact assessment can be one of the outputs of a BIM-based design process, it must be noted that to date the applications of this tool to modelling already existing buildings are still relatively uncommon, and that in general BIM can be useful in those cases where large, complex projects are to be designed, built and managed by a number of stakeholders, rather than in smaller, owner-controlled ones such as those at Sieben Linden.

5.2 Sidenotes on numerical findings

In 2014, the territorial extension of Sieben Linden land was 82.5 ha, which corresponded to a biocapacity of 434.1 gha (3.42 gha/person). Although methodologically incorrect – as the yield factors used to calculate the EF and biocapacity are the average global one in the first case and the average national one in the second – it might be interesting to compare this figure with Sieben Linden's EF, which was 391.16 gha (3.08 gha/person).

The actual "public services" EF of Sieben Linden residents must be lower than the per capita share flatly applied in this calculation. In spite of this, we believe that no differentiation between citizens of a certain country should be made in this respect, as it would be socially unfair, and anyhow extremely difficult to calculate. However, it should be remarked that in some areas Sieben Linden as a community – or its individual residents in their everyday lifestyle – are less dependent on public services than the average German citizen. For instance, we are referring here to infrastructure such as faeces treatment plants or kindergartens, which the community does not need as they provide such services by themselves.

Some further EF might be hidden in some cases. For instance, the remarkable (and growing) amount of time spent by Sieben Linden residents on the Internet does not only imply a local energy consumption, but also an extra remote consumption where the servers are located and all along the service-providing infrastructure (data transmission energy, which depends on its speed). We have data (resulting from a limited number of interviews and therefore not necessarily representative) on the average time spent on the Internet (see Chapter 2.4.2), but we do not know how such time was used. In Sieben Linden, the use of mobile phones and the wi-fi transmission of data are avoided for hygienic reasons; people must connect their devices to the web via cable. This is very positive both from the point of view of health and that of energy consumption. However, to attempt an estimate of the hidden electrical consumption it would be necessary to know also what kind of data are transferred/exchanged; it is for example known that streaming video is the most demanding activity (Shehabi et al. 2014). De Decker (2015) affirms that "the energy use of the internet can only stop growing when energy sources run out, unless we impose self-chosen limits, similar to those for cars (. . .). Limiting demand would also imply that some online activities move back to the off-line world" (See also Hilty 2008).

Another example of hidden EF might be found in the growing tendency to buy goods from the web: it is doubtful whether the delivery of such goods has the same impact on the environment as shopping the traditional way.

5.3 Recommendations

Generally speaking, recommendations regarding possible policies to be adopted in order to decrease Sieben Linden's EF are hard to formulate as they may depend on the environmental metrics adopted: a carbon-only as well as a carbon-plus-energy analysis would indicate different weaknesses and strengths than an EF assessment.

First of all, we would like to remark that in spite of the conspicuous share of the EF that is due to burning firewood, this should not disquiet as this is obtained sustainably, in the strict sense of the word (see Chapter 2.4.1); paradoxically, if all the houses in Sieben Linden ran on nuclear energy rather than firewood, there would be an instant 15% drop in the total EF, but this would be not much good for the planet. (Environmental impact assessment methodologies based on carbon often make the impact of burning firewood = 0.)

This said, there are areas that might be addressed by impact reduction efforts. Travel patterns might be collectively discussed to find further economies in distances run yearly, and to optimise the private cars' occupancy, that is, to extend carpooling practices. (Of car travel 193,949 km are shared in a year – that is, 36% of the global distance run by car. The mean car occupancy rate is 2.27 persons.) Obviously a remarkable reduction of travel EF would be achieved if Deutsche Bahn restarted operating the local railway, presently substituted by a bus service.

Self-sufficiency is not an absolute goal in terms of sustainability. However, there seems to be scope for increasing self-sufficiency at least in some areas, for instance installing a wind generator.

Trends towards a certain "normalisation" of practices are manifesting, such as, for instance, the increase in the possession and use of personal electronics; some mechanisation of the organic gardening activity (Langkabel in Stanellé 2017:97 ff.); and the imposed connection to the water network: as long as possible, such tendencies should be attentively contained.

In Sieben Linden, there is no scope for giving priority to retrofitting buildings as very few constructions pre-date the settlement of the community. (Of course, it remains a patent contradiction to establish a low-environmental-impact village by building a new settlement, if compared to refurbishing an existing one: but it is almost impossible to acquire a whole abandoned village where to start such an experiment – both because of lack of availability and because of property price. One remarkable exception is Lebensgarten Steyerberg, in Lower Saxony, which was born out of absolutely unique circumstances.)

However, it remains questionable whether living in trailers (a "housing" solution relatively widespread in ecovillages, which recalls their often alternative-lifestyle origins) is as viable an option as in buildings: on the one hand, improving the thermal resistance of a trailer would entail much lower EE and EC than building a new house; on the other, no reliable information is today available as regards the trailers' firewood consumption (per square metre and per inhabitant), so to compare it with that of the houses. It is therefore recommended to monitor firewood consumption for each building and each trailer (or group of trailers), not in bulk as current practice, so to help understanding the real consumption patterns in detail.

Moreover, it might be relevant – scientifically but also practically – to measure in detail material and energy flows on a building site on a daily basis, in order to portray a whole and reliable picture of EE and EC associated to new construction (and therefore cover phases A4 and A5 of LCIA, see Chapter 3.6). Analyses are more accurate when foreseen since the design phase and are conducted during all of the construction process, and fully integrated in it. Another possible research might be dedicated to the detailed investigation of trailers' impacts – both in the construction and in the operation phases. Such research is furthermore relevant, as today not only second-hand caravans are in use in Sieben Linden, but also new ones built in the ecovillage itself. In this specific field then, there would be much scope for an ecological impact comparison between upgrading and new construction.

The problems encountered in comparing the EE and EC values obtained for Sieben Linden buildings to others taken from the scarce literature on the subject (see Chapter 3.5.2) make it very difficult to draw conclusions regarding the way Libelle fares in comparison with an average timber building. In spite of this, our analysis suggests that the "villa Strohbunt model" is actually the only one which can guarantee a radical reduction of a building construction's footprint. On the other hand, the "Libelle model" offers houses with a lower use-life impact (because of the professional execution, but above all because of the lower energy requirement). Future experimentation should therefore try to merge the positive aspects of the two models in new buildings.

5.4 Sieben Linden as a living laboratory of sustainability

This study of ours obviously belongs to a research field much concerned with climate change. Yet, we are aware that the current environmental crisis does not lie in our carbon footprint only. The issue is first of all political and cultural: The real question is not much about envisaging less polluting technologies, rather it is about deciding whether to insist on the present "development (i.e., growth) model" or to decide to switch to a new paradigm (see, among others, Capra 1996). It is exactly in this field that some successful ecovillages such as Sieben Linden have strong models and solutions to offer: the societal and cultural innovation they bring forth is worthwhile because of their experimentation with alternative value frames and ways of living a modern life, decoupling wellbeing and fulfilment from high resource consumption and environmental pollution.

Sieben Linden may be indicated as a living laboratory where a special lifestyle is experimented, which allows (among other things) to recover some attitudes that went lost in the modern world, particularly in terms of sharing, community, self-help, but also in terms of dependence from local environmental resources, sobriety and reusing practices. Just to make an example, the very existence of a "second-hand goods exchange corner" (a facility located at the very centre of the village, which everybody seems to use) is an obvious incentive to checking if something you are looking for is readily available, before even considering the buy option.

This research shows the extent of the potential for environment-friendly societal innovation which can be acted by a small community such as Sieben Linden. Kunze (2016) has documented and discussed this from a different, and actually complementary, point of view. Societal innovation covers different dimensions and scales, from the very establishment of a thorough and remarkably consistent self-regulating community (made up of a number of formal associations, neighbourhood groups, etc., each one with specific tasks and spheres of action, see Chapter 1.2) to the re-crafting of everyday practices. Many of the preliminary rules of the ecovillage bear recognisable environmental consequences, such as avoiding private ownership of houses and land, banning motor vehicles from within the settlement (one of the positive ramifications is that no street lighting is needed), banning domestic animals (the ban includes both pets with their high environmental impact (Vale 2009) and livestock, even hens: this creates some agricultural management issues and also some contradictions, since eggs need to be bought), opting for radically organic farming, taking a vegetarian/vegan diet in collective meals (which are not compulsory but are chosen by the majority of residents), treating greywater in a reed bed, setting up compost toilets only, etc. (see Chapter 1.5).

None of these (and other) in-group stipulations imposes particularly harsh sacrifices; residents wilfully accept them as part of their decision to take an ecovillage lifestyle (although some of our interviewees seemed constrained by the inevitability of limiting visits to family or friends, or seemed to feel guilty for travelling) – and this is in tune with some contemporary tendency towards a "3rd millennium monastery" model (Pallante 2013). Referring to future "sustainability," English ecologist Edward Goldsmith actually said: "if I had to pass on only three ideas, I would say: a vegetable garden, a community and faith." Sieben Lindeners are not a community united by a religious faith – they are, in fact, "a village without a church" (Würfel 2012). Still, they share the acceptance of taking an utterly unusual degree of awareness and responsibility in their actions; what we would call the secular belief in (and the engagement towards) a simpler life, that is less impactful on the environment, and nevertheless rich at least socially and culturally if not spiritually. Such "faith" is complemented by a strong engagement towards the building and the careful maintenance of themselves as a community, and towards the piecemeal implementation of a plan for the physical development of the settlement which encompasses buildings, facilities, vegetable gardens, arable land, forest, etc. (see Chapter 1.2) – they seem therefore to meet Goldsmith's requirements for minimum survival kit.

It is relevant to remark that social innovation developed in Sieben Linden does not include only "don'ts" (things that are commonly performed in the society at large and are unusual or downright avoided here), but also leaves a lot of space to possibilities which are not practised or downright impossible somewhere else. For instance, the extraordinary situation – both in time, a few years after the German reunification; and in space, in a depopulated, marginal area of former German Democratic Republic – in which the ecovillage started allowed room for experimentation and development of innovations in the field of building: straw bale building in Germany started from Sieben Linden, and this is where FASBA was first established (see Chapter 1.8). Sieben Linden offers also extraordinarily numerous leisure and cultural/social opportunities (e.g. dance, cinema, sauna . . .) compared to the smallness of the village. This creates the advantage of minimising the reasons for travel for residents (the obvious other side of the coin is the potential reduction of interaction with neighbouring society).

The case of Sieben Linden also shows that even a small group of people can have an impact on public policies if it behaves coherently. For instance, residents and guests generate so high a demand for public transportation that the local bus line is the busiest rural bus line in the whole state of Saxony-Anhalt, and the ecovillage has become one of the stakeholders in the discussion about local transportation network design.

As we saw in Chapter 1.1, Sieben Linden was born thanks to an exception to the law on land use – a new settlement area was earmarked out of agricultural land. This makes it very peculiar to frame the Sieben Linden case in "urban development" or "rural development" narratives, as commonly understood. It is also questionable if any finding here can be directly applied to an urban neighbourhood or rural village, even in the not-so-likely scenario of a locally shared willingness to fulfil Sustainable Development Goals. Yet, we are convinced that the experimentations conducted at such a large scale can be useful, *mutatis mutandis*, at a smaller one, with Sieben Linden playing both the role of demonstration piece and of initiator of local processes (see Chapter 5.6). Kunze (2016) aptly describes social innovation in Sieben Linden as a continuous process of 1) experimentation; and 2) stabilisation of the appropriate findings into professionalised/institutionalised practices. Some of the latter are of course born from local needs and desires, but can to some extent be adapted and adopted by the "society at large."

Similarly, we believe the methodology we used in this analysis can be fruitfully applied to virtually any small town (just to make an example: transition towns, etc.), also because the mathematic implied in it is elementary and no elaborate, proprietary software has been used to perform the calculations (a few spreadsheets are all one needs . . . provided that input data are available). We insist on the relatively small dimension of the subject observed, not only for obvious practical reasons, but much more so in the conviction that only if communities are small enough they can effectively control the use of resources within their own territory.

5.5 Structural limitations

The Vales show that to stay within the "fair share" a very strong reduction of the footprint would be needed in industrialised countries (Vale 2013). In the same book, and perhaps not so incidentally, Rees and Moore (Rees 2013:19 ff.) found useful to compare the EF of Vancouver with that of two ecovillages, and used the latter as a benchmark to show that the extent of a possible reduction of the personal footprint would be about two thirds. Such reduction could only be attained through radical changes in the energy intensity of food production and processing, transport, building use, etc.: for instance, "i) a 50 per cent reduction in private vehicle use coupled with elimination of air travel, or ii) elimination of virtually all private vehicles" (Rees 2013:23).

What is more, it is still to be observed that, according to Bjørn and Hauschild (2015), the "absolute environmental sustainability" threshold for keeping the climate change within an increase of 2°C would be 985 $kgCO_{2eq}$ per capita yearly (if flatly shared between the global population), less than half of the average Sieben Linden value of 2,430 $kgCO_{2eq}$/person*year, public services excluded (see Chapter 2.5.1). There is, therefore, scope for a further reduction of Sieben Linden's environmental impact (see Chapter 5.3), but that would by no means be enough. In fact, for all the commendable (and successful, if compared with average values) concern the Sieben Linden community may have in front of limiting their EF, this research shows that it would be impossible to stay within the "fair share" unless a drastic reduction of the EF associated to public services, government, etc., is enforced. Such spheres lay far outside of the control of an individual or a group of citizens: it is therefore at the political level that more courageous steps towards sustainability should be taken.

Not only "energy efficiency and decarbonisation policies need to be combined with [behavioural change policies – be them based on price and information, or on attitudes and values –] if we want energy use and carbon emissions to go down." Unfortunately, "two decades of climate-change related awareness campaigns have not [succeeded in decreasing] energy demand and carbon emissions." According to De Decker (2018) – whose arguments are much indebted to those of Shove (2010 and Shove et al. 2012; see also Southerton et al. 2011), this is due to the fact that, contrary to common assumption, it holds not true that what people do is in essence a matter of individual choice. (. . .) Obviously, individuals do make choices about what they do and some of these are based on values and attitudes. For example, some people don't eat meat, while others don't drive cars, and still others live entirely off-the-grid. However, the fact that most people do eat meat, do drive cars, and are connected to the electric grid is not simply an isolated matter of choice. (. . .) What people do is also conditioned, facilitated and constrained by societal norms, political institutions, public policies, infrastructures, technologies, markets and culture.

De Decker goes on arguing that

> by placing responsibility – and guilt – squarely on the individuals, attention is deflected away from the many institutions involved in structuring possible courses of action, and in making some very much more likely than others. (. . .) Focusing on individual responsibility is in line with neoliberalism and often serves to suppress a systemic critique of political, economic and technological arrangements.

On one hand, then, a bold and systemic change in priorities in the allocation of resources, the way public services are provided, and the way the infrastructures and institutions are arranged should be made, that is consistent with the numerous international "goals," "targets," "protocols" etc. that so lightheartedly governments have been subscribing in the last decades – one that would be less impactful in itself, and would make easier for individuals to opt for less impactful choices.

On the other hand, De Decker proposes that much more attention should be put on "the socially embedded underpinning of behaviour," that is,

> the social organisation of everyday practices such as cooking, washing, shopping, or playing sports. (. . .) Social change is about transforming what counts as 'normal' – as in smoke-free pubs or wearing seat belts. (. . .) A systemic approach to sustainability encourages us to imagine what the 'new normal' of everyday sustainability might look like.

(see also Spurling et al. 2013).

Until the socio-economic-political system does not take a radical attitude towards "sustainability," welcome are those sub-systems that are experimenting (in so far as they can) with the way everyday behaviours are conducted (establishing attempts of "new normal" ways of doing things) as well as with collective measures that would be much more effective if taken at a higher level.

It would be very interesting to see a "natural" (as opposed to an "intentional") community arising, which would bring forward structured, coherent, clear, and shared instances as in the case of Sieben Linden, even though non-cogent and gradually implemented. But what local social body – if not just an intentional community, be it an ecovillage or a monastery – has the strength to pursue such a path for a long time? Actually, it is already a miracle if rural communities can provide themselves with a plan or pattern to survive, and to implement them successfully, as for instance in the cases of Kamiyama (Tokushima prefecture, Japan), and the area west of Plauer lake (Mecklenburg, Germany) (Yoshimoto 2017; Frech et al. 2017).

In the context of sustainable rural development, it is also noteworthy that Sieben Linden is one of the partners of a project, called "Leben in zukunftsfähigen Dörfern" ("Living in Sustainable Villages"), launched in April 2017 and now close to completion. This project is funded by the UBA, and puts five member communities of GEN Germany from five federal states (Saxony-Anhalt, Thuringia, Baden-Württemberg, Lower Saxony and Hessen) together with some natural villages selected from their surroundings. The goal is to design sustainable rural development concepts as well as concrete holistic initiatives, seeing if ecovillages can act as change catalysts, transferring solutions in the ecological, economic, social and cultural dimensions of sustainability, or making their inspirational power available to the surrounding communities to help tackle their current problems (such as, *inter alia*, emigration and ageing, loss of cultural landscape and diversity as well as social and cultural stagnation) (GEN Deutschland 2018).

This very promising initiative shows that although nested in a rather marginal niche away from big cities, Sieben Linden is not an "external" "alternative," "opposed" to the city, nor it was born from an intention of retreat and escapism from the "outside world." True, the country is where people go in times of crises – this is where most monasteries have been established in history – and the physical environment of Sieben Linden is very sparse. Yet, the ecovillage is anything but closed; it is on the contrary densely interconnected with its territorial context at various levels. To give a few examples (see Chapter 1.8): its economy, albeit self-enclosed to a nowadays uncommon level, is obviously integrated in the local, national and to a global economy; it networks and cooperates with local administrations and entities; it hosts seminars and courses, attended by thousands of people each year nationwide; it established itself as an observable experiment (the number of studies and dissertations grows at a faster rate than the number of residents); and it has ever been playing the role of communication hub for European ecovillages. (One more obvious evidence of the fact that the "Sieben Linden utopia" has nothing to do with some current trend towards a "re-ruralisation" hinted with nostalgic or neo-conservative tones (Voigts 2016) is that very few of the residents are engaged in agricultural or forestry activities.)

In conclusion, to come back to the environmental impact issues, the "Sieben Linden experiment" shows very clearly which are the areas open to a small group of committed people's agency and which cannot be dealt with efficiently except at a higher political level. In our opinion, this is particularly relevant as governments are trying to figure out which policies to implement in order to meet the Sustainable Development Goals they have subscribed (SDG 2015).

5.6 Future work paths

We believe the establishment of a very basic "observatory" of environmental indicators (such as the data collected to carry this study on) would be very helpful to a community as committed as Sieben Linden is. Such a permanent observatory might be equipped with a calculation tool where data analysed in this study may be yearly inputted to obtain a simplified Sieben Linden EF assessment in the future.

On one hand, future work could progress in merging EF and LCA methodologies in order to achieve an absolute sustainability assessment based on a bottom-up approach (that is, on the component method).

On the other hand, our hope is to be able to progress with the investigation covering areas we could not include in this study, which are related to a broadly understood concept of sustainability, albeit not strictly connected to EF assessment. In fact, so far we have focused on negative environmental impacts accounting. In the present study we extensively referred to reference methodologies such as the Carbon Footprint (Wiedmann 2008) and the Ecological Footprint (Wackernagel 1996). These methodologies are extensively found in literature and are mostly used in order to see where to take action to minimise the damage to the environment and human health, and to use resources more efficiently.

However, regenerative design practices aim not only to minimise negative environmental impacts, but also to support the social and natural capital, and to increase its health. Important supporting theories derive from the area of sustainable design, but also include ecology and agriculture. Some of the most significant are John Tillman Lyle's Regenerative Design (Lyle 1994), Bill Reed's work with Regenesis Group (Mang 2012), John and Nancy Todd's Ecological Design (Todd 1984), Janis Birkeland's Positive Development (Vandenbroeck 2010), Jason F. McLennan's Living Community Challenge (Thomas 2016) and Bill Mollison's Permaculture (Mollison 1991). Most of these theories are discussed by Hes and Du Plessis (2014).

Other relevant methodologies can be borrowed from Landscape Ecology, especially for what concerns the recovery of degraded landscapes and biodiversity protection. Classification and mapping of ecosystem services

(ES) may provide significant information about the interactions between humans and nature; highlighting ES variations can point out the efficacy of regenerative practices.

Our aim would be to evaluate regenerative practices and the positive impacts that human actions may have on the natural environment and on society: such evaluation will obviously integrate a wide variety of impacts, merging the results of sectorial assessments.

Ecovillages like Sieben Linden appear to be a promising investigation field since they aim to create healthy and equitable living environments minimising impacts on nature. For instance, it is evident even to the layperson's eye that the amount of biodiversity that can be found within the community's land is much higher than that of the surrounding agricultural land – a landscape heavily disrupted by extensive monoculture.

We are convinced that such a methodology might ideally be useful to support decision-making by communities and local governments in sustainable design, and promote regenerative practices in rural and urban areas, in the framework of Goals #11, 12, 13 and 15 of the 2030 Sustainable Development Agenda.

References

Bjørn, Anders; Michael Zwicky, Hauschild, "Introducing carrying capacity-based normalisation in LCA: framework and development of references at midpoint level", *International Journal of Life Cycle Assessment*, 20, 2015, pp. 1005–1018.

Capper, Graham; Jane Matthews; Steve Lockley, "Incorporating embodied energy in the BIM process", CIBSE ASHRAE Technical Symposium, Imperial College, London, April 18–19, 2012.

Capra, Fritjof, *The Web of Life. A New Synthesis of Mind and Matter*, London: Harper Collins, 1996.

Castellani, Valentina; Serenella Sala, "Ecological footprint and life cycle assessment in the sustainability assessment of tourism activities", *Ecological Indicators*, 16, 2012, pp. 135–147. doi:10.1016/j.ecolind.2011.08.002

Chau, C.K.; T.M. Leung; W.Y. Ng, "A review on Life Cycle Assessment, Life Cycle Energy Assessment and Life Cycle Carbon Emissions Assessment on buildings", *Applied Energy*, 143, April 1, 2015, pp. 395–413. doi:10.1016/j.apenergy.2015.01.023

De Decker, Kris, "Why we need a speed limit for the internet", *Low-Tech Magazine*, October 19, 2015 [online]. Available from: www.lowtechmagazine.com/2015/10/can-the-internet-run-on-renewable-energy.html [last viewed Aug. 2018].

De Decker, Kris, "We can't do it ourselves", *Low-Tech Magazine*, July 5, 2018 [online]. Available from: www.lowtechmagazine.com/2018/07/we-cant-do-it-ourselves.html [last viewed Aug. 2018].

Frech, Siri; Babette Seurell; Andreas Willisch (eds.), *Neu Land gewinnen. Die Zukunft in Ostdeutschland gestalten*, Berlin: Ch-Links Verlag, 2017, pp. 126–135.

GEN Deutschland, *Leben in zukunftsfähigen Dörfern*. Available from: gen-deutschland.de/wp_gen/projekte/uba-projekt/ [last viewed Sep. 2018].

Hes, Dominique; Chrisna Du Plessis, *Designing for Hope: Pathways to Regenerative Sustainability*, London: Earthscan, 2014.

Hilty, Lorenz M., *Information Technology and Sustainability: Essays on the Relationship Between ICT and Sustainable Development*, Norderstedt: Books on Demand, 2008.

König, Holger et al., *A Life Cycle Approach to Buildings: Principles Calculations Design Tools*, München: Institut für internationale Architektur-Dokumentation, 2010.

Kunze, Iris; Sabine Hielscher, *Fallstudienbericht COSIMA: Entwicklung der Klimaschutzinitiativen*, Poppau: Ökodorf Sieben Linden, 2016.

Lewan, Lillemor; Craig Simmons, *The Use of Ecological Footprint and Biocapacity Analyses as Sustainability Indicators for Sub-national Geographical Areas: A Recommended Way Forward*. Final Report, Prepared for Ambiente Italia ECPI (European Common Indicators Project), 2001. Available from: http://manifestinfo.net/susdev/01EUfootprint.pdf [last viewed Sep. 2018].

Lyle, John T., *Regenerative Design for Sustainable Development*, Brisbane: Wiley and Sons, 1994.

Mang, Pamela; Bill Reed, *Regenerative Development and Design*, New York: Springer Encyclopedia of Sustainability Science and Technology, 2012.

Mollison, Bill; Reny Mia Slay; Andrew Jeeves, *Introduction to Permaculture*, Erskineville: Tagari publications, 1991.

Oekobaudat Informationsportal Nachhaltiges Bauen, *ÖKOBAUDAT 2017-I – EN 15804 und BNB-konforme Daten für über 1000 verschiedene Bauprodukte*, Bundesministerium des Innern, für Bau und Heimat, November 27, 2017. Available at: www.oekobaudat.de [last viewed Sep. 2018].

Pallante, Maurizio, *Monasteri del terzo millennio*, Torino: Lindau, 2013.

Pomponi, Francesco; Alice Moncaster, "Embodied carbon mitigation and reduction in the built environment – What does the evidence say?", *Journal of Environmental Management*, 181, October 1, 2016, pp. 687–700. doi:10.1016/j.jenvman.2016.08.036

Rees, William E.; Jennie Moore, "Ecological footprints, fair earth-shares and urbanization", in Robert Vale; Brenda Vale (eds.), *Living Within a Fair Share Ecological Footprint*, Abingdon: Routledge 2013, pp. 3–32.

SDG, *Sustainable Development Knowledge Platform*, 2015 [online]. Available from: https://sustainabledevelopment.un.org/about [last viewed Sept. 2018].

Shehabi, Arman; Ben Walker; Eric Masanet, "The energy and greenhouse-gas implications of internet video streaming in the United States", *Environmental Research Letters*, 9, 2014.

Shove, Elizabeth, "Beyond the ABC: Climate change policy and theories of social change", *Environment and Planning A*, 42, 6, 2010, pp. 1273–1285.

Shove, Elizabeth; Mika Pantzar; Matt Watson, *The Dynamics of Social Practice: Everyday Life and How It Changes*, London: Sage, 2012.

Southerton, Dale; Andrew McMeekin; David Evans, *International Review of Behaviour Change Initiatives: Climate Change Behaviours Research Programme*, Edinburgh: Scottish Government Social Research, 2011.

Spurling, Nicola Jane et al., *Interventions in Practice: Reframing Policy Approaches to Consumer Behaviour*, Sustainable Practices Research Group Report, September 2013.

Stanellé, Chironya; Iris Kunze (eds.), *20 Jahre Ökodorf Sieben Linden*, Poppau: Freundeskreis Ökodorf, 2017.

Thomas, Mary A., *The Living Building Challenge: Roots and Rise of the World's Greenest Standard*, Seattle: Ecotone Publishing, 2016.

Todd, Nancy J.; John Todd, *Bioshelters, Ocean Arks, City Farming: Ecology as the Basis of Design*, San Francisco: Sierra Club Books, 1984.

Vale, Robert; Brenda Vale, *Time to Eat the Dog? The Real Guide to Sustainable Living*, London: Thames and Hudson 2009.

Vale, Robert; Brenda Vale (eds.), *Living Within a Fair Share Ecological Footprint*, Abingdon: Routledge 2013.

Vandenbroeck, Philippe, *Janis Birkeland's 'Positive Development'. A Strategy Towards a Sustainable Built Environment*, 2010. Available from: www.shiftn. com/media/SN_RP_janisbirkeland_v02LOW.pdf [last viewed Sept. 2018].

Voigts, Eckart, "LandLust – The 'knowability' of post-pastoral ruralism", in Vanessa Miriam Carlow; Institute for Sustainable Urbanism ISU (eds.), *Ruralism: The Future of Villages and Small Towns in an Urbanizing World*, Berlin: Jovis, 2016, pp. 162–178.

Wackernagel, Mathis; William Rees, *Our Ecological Footprint: Reducing Human Impact on the Earth*, Gabriola Island: New Society Publishers, 1996.

Wiedmann, Thomas; John Barrett, "A review of the ecological footprint indicator. Perceptions and methods", *Sustainability*, 2, 2010, pp. 1645–1693. doi:10.3390/su2061645

Wiedmann, Thomas; Jan Minx, "A definition of 'carbon footprint'", in C. C. Pertsova (ed.), *Ecological Economics Research Trends*, Hauppauge, NY: Nova Science Publishers, 2008, pp. 186–197.

Woolley, Tom, *Low Impact Building: Housing Using Renewable Materials*, Chichester: Wiley-Blackwell, 2013.

Würfel, Michael, *Dorf ohne Kirche. Die ganz grosse Führung durch das Ökodorf Sieben Linden*, Poppau: Eurotopia-Buchversand, 2012.

Yoshimoto, Mitsuhiro, "Kamiyama's success in creative depopulation", *Field. A Journal of Socially-Engaged Art Criticism*, 8, Fall 2017.

Index

absolute environmental sustainability 2, 90
agency 93
agriculture 24, 81; *see also* organic cultivation
Altmark 5–6, 10

Beetzendorf 5, 23
behaviour 90–91
biocapacity 29, 39, 49, 81–82, 84
biodiversity 1, 81, 93–94
biologically productive land 28–29
bioproductive capacity 29, 81
Birkeland, Janis 93
Bjørn, Anders 2, 90
boundaries 27, 30, 42, 56, 69, 70–71
Brunnenwiese 20
buildable area 12, 18
building: materials 14, 56, 71, 83; site 56, 70, 84; technique 11, 18
Building information modeling (BIM) 84
businesses 28

camping 12, 15
caravan(s) 15, 19, 20–21, 87; *see also* trailer(s)
carbon dioxide 27–28, 55, 57, 82; *see also* Global warming potential (GWP)
Carbon Footprint (CF) 2, 27–28, 40, 55
Carbon Trust 28, 50
children 12, 15; *see also* kindergarten
circular economy 6, 10
climate change 27, 87, 90
Club99 9, 17–18, 56

community: building 7, 16, 22; facilities 12; kitchen 7, 12, 32; living 6; spaces 22
component method 29–30, 71, 93; *see also* compound method
compost(ing) toilets 12, 14, 88; *see also* (faeces) treatment plant
compound method 29; *see also* component method
cradle to gate phase 2, 57, 71

Dangelmeyer, Peter 1–2, 27, 42; *see also* University of Kassel
databases 1, 70, 83
data libraries 2; *see also* databases
decision-making 94
De Decker, Kris 85, 90–91
Deutsche Bahn 86
Deutsche Bundesstiftung Umwelt (DBU) 6, 24
diet 11, 12, 88; *see also* food
Du Plessis, Chrisna 93

Ecological Footprint (EF) 2, 27–28, 45, 71, 81
ecological footprint standards 29; *see also* Global Footprint Network (GFN)
economy 9, 92; *see also* circular economy
ecosystem service(s) 81, 93
EcoTransit 38
Embodied carbon 55, 69; *see also* Global Warming Potential (GWP)
Embodied energy 38, 55, 58, 61, 67, 69; *see also* Primary energy intensity (PEI)

Environmental impact 3, 27–28, 81, 83–84, 86, 93
EUREAPA 39, 41, 46, 50, 76
Eurotopia 24

Fachverband Strohballenbau Deutschland e.v. (FASBA) 24, 89
fair share [Ecological Footprint] 47, 90; *see also* Vale, Robert and Brenda
farmland 21
firewood 14, 32, 43, 82; *see also* heating
food 10, 12, 37–38, 40, 45; *see also* diet
forest 14, 32, 33, 56, 58; *see also* firewood
Freie Schule Altmark e.v. 10; *see also* kindergarten
Freiwillige Ökologische Jahr (FÖJ) 15–16, 23, 31
Freundeskreis Ökodorf e.v. 8

Global Ecovillage Network (GEN) 1, 5, 24
Global Footprint Network (GFN) 29, 38, 39, 46–47
Global warming potential (GWP) 55, 58–59, 61, 77; *see also* carbon dioxide
Globolo 12, 16, 19
Goldsmith, Edward 88
greywater 12, 21, 62
guests 11, 15, 23, 31

Hauschild, Michael Zwicky 90
heating 11, 32, 61; *see also* firewood
Hes, Dominique 93
Housing cooperative 6, 8, 10; *see also* Wohnungsgenossenschaft Sieben Linden e.G. (WoGe)

indicators 2, 93
individual responsibility 91
infrastructure 39, 76, 85
intentional community 92
interviews 22, 31, 33, 85
inventorying 84
Inventory of Carbon and Energy (ICE) 56–58, 82–83; *see also* database

jobs 8, 10

Kamiyama 92
kindergarten 12, 14–15; *see also* Freie Schule Altmark e.v.
Kranich 21
Kunze, Iris 5, 8, 11, 88, 89

landscape ecology 93
Lebensgarten Steyerberg 86
Lewan, Lillemor 29–30, 49
Libelle 19, 21, 55–66, 68–69, 71–72, 75, 77, 79, 87
Life cycle assessment (LCA) 2, 28, 33, 58, 61–62, 83, 93
lifestyle 1, 5, 18, 27, 87; *see also* behaviour
Lyle, John Tillman 93

McLennan, Jason F. 93
mobility 11–12, 81
Mollison, Bill 93
motor vehicles 12, 37, 86; *see also* private, car; private, vehicle

Nachtigall 21
National footprint account 29; *see also* ecological footprint standards
Naturwaren Sieben Linden e.v. 9–10
neighbourhoods 9, 12, 17, 19
Nordhaus 7, 17, 22
Nordriegel 15–17, 22
Nordschonung 17–18

Ökobaudat 83; *see also* database
operation(al): emissions 77–79; energy 12, 77–78; phase 87
organic cultivation 12

panels: PV 11, 32; solar 12, 21, 32–33, 44, 61; *see also* solar systems
permaculture 7
personal electronics 86
Plauer lake 92
policy 85, 89–91, 93; *see also* decision-making
Primary energy intensity (PEI) 2, 55, 58, 61, 69, 77; *see also* Embodied energy

Private: apartment 22; car 12, 37, 86; kitchen 20, 32; property 8; space 11; vehicle 90; *see also* motor vehicles
public services 49, 75, 85, 90–91
public transport(ation) 6, 37, 89

railway 7, 86; *see also* train
Reed, Bill 93
reed bed 7, 12, 21, 62, 88; *see also* (feaces) treatment plant
Rees, William 28, 90
Regiohaus 7, 12, 16–17, 22
regional development 5
resources: consumption of 30; regional 18; renewable 29, 72
retrofitting 86
Rohkostversand "Raw Living" 10
rural development 89, 92

sauna 8, 20
Saxony-Anhalt 5, 7, 10, 89, 92
self-sufficiency 5, 8, 10, 86
seminars 7–10, 12, 17, 23, 31, 92
service life 2, 75–76
Settlement cooperative 6, 8, 10; *see also* Siedlungsgenossenschaft Ökodorf e.G. (SiGe)
sharing 9, 12, 18, 20, 47, 87
Shove, Elizabeth 90
Siedlungsgenossenschaft Ökodorf e.G. (SiGe) 1, 8, 21; *see also* Settlement cooperative
site 6–7, 14, 21, 55–56
societal innovation 3, 88–89
solar systems 11, 62; *see also* panels
Sommer, Jörg 5–6
Sonneneck 7, 16–17
spatial planning 6

straw (bales) 14–15, 18–20, 23–24, 56, 65, 71, 84, 89
Strohpolis 19
Strünke, Christoph 8–9, 12, 23
Südhaus 7, 17, 22
Sustainable Development Agenda 94
Sustainable development goal(s) (SDG) 1, 3, 89, 93

TAT-Orte-Preis 6
Todd, John and Nancy 93
trailer(s) 8, 12, 19, 22, 39, 43, 86–87; *see also* caravan(s)
train 37; *see also* railway
(faeces) treatment plant 35, 85; *see also* compost(ing) toilets

University of Kassel 1, 41

Vale, Robert and Brenda 47, 88, 90
Villa Strohbunt 18, 22, 55–61, 65–69, 71–72, 77, 79

Wackernagel, Mathis 28–29, 93
water 11, 17–18, 28, 32, 43, 61–62, 86
Wegmann-Gasser house 65, 69, 72, 83
Wiedmann, Thomas 27, 81, 93
Windrose 19, 20, 21
Wohnungsgenossenschaft Sieben Linden e.G. (WoGe) 8, 19, 20; *see also* Housing cooperative
workshop: carpenter's 14; 12; straw bale 15; theoretical and practical 23; *see also* seminars
Würfel, Michael 12, 88

yield factor 29, 82, 84
youth 9, 12, 15